Post-socialist Cities and the Urban Common Good

This book explores the changing approaches to urban common good in Central and Eastern Europe after 1989. The question of common good is fundamental to urban living; however, understanding of the term varies depending on local contexts and conditions, particularly complex in countries with experience of communism.

In cities east of the former Iron Curtain, the once ideologically imposed principle of common good became gradually devalued throughout the 20th century due to the lack of citizen agency, only to reappear as a response to the ills of neoliberal capitalism around the 2010s. The book reveals how the idea of urban common good has been reconstructed and practiced in European cities after socialism. It documents the paradigm shift from city as a communal infrastructure to city as a commodity, which lately has been challenged by the approach to city as a commons. These transformations have been traced and analysed within several urban themes: housing, public transport, green infrastructure, public space, urban regeneration, and spatial justice. A special focus is on the changes in the public discourse in Poland and the perspectives of key urban stakeholders in three case-study cities of Gdańsk, Kraków, and Łódź. The findings point to the need for drawing from best practices of the socialist legacy, with its celebration of the common. At the same time, they call for learning from the mistakes of the recent past, in which the opportunity for citizen empowerment has been unseized.

The book is intended for researchers, academics, and postgraduates, as well as practitioners and anyone interested in rediscovering the inherent potential of urban commonality. It will appeal to those working in human geography, spatial planning, and other areas of urban studies.

Maja Grabkowska is a human geographer and Assistant Professor at the Department of Socio-Economic Geography, University of Gdańsk, Poland. She authored and co-authored research publications on post-socialist urban regeneration, gentrification, and grassroot initiatives, including a book titled *Regeneration of the Post-socialist Inner City: Social Change and Bottom-up Transformations in Gdańsk* (2012). She is also a co-founder of Sopocka Inicjatywa Rozwojowa, an informal citizen group in Sopot, Poland, advocating participatory democracy and sustainable development at the local level.

Routledge Contemporary Perspectives on Urban Growth, Innovation and Change

Series edited by Sharmistha Bagchi-Sen, Professor, Department of Geography and Department of Global Gender and Sexuality Studies, State University of New York-Buffalo, Buffalo, NY, USA, and **Waldemar Cudny**, Associate Professor, working at the Jan Kochanowski University (JKU) in Kielce, Poland.

Urban transformation affects various aspects of the physical, social, and economic spaces. This series contains monographs and edited collections that provide theoretically informed and interdisciplinary insights on the factors, patterns, processes, and outcomes that facilitate or hinder urban development and transformation. Books within the series offer international and comparative perspectives from cities around the world, exploring how 'new life' may be brought to cities, and what the cities of future may look like.

Topics within the series may include: urban immigration and management, gender, sustainability and eco-cities, smart cities, technological developments and the impact on industry and on urban societies, cultural production and consumption in cities (including tourism, events and festivals), the marketing and branding of cities, and the role of various actors and policy makers in the planning and management of changing urban spaces.

If you are interested in submitting a proposal to the series, please contact Faye Leerink, Commissioning Editor, faye.leerink@tandf.co.uk.

The Interstitial Spaces of Urban Sprawl
Geographies of Santiago de Chile's *Zwischenstadt*
Cristian A. Silva

Mega Events, Urban Transformations and Social Citizenship
A Multi-Disciplinary Analysis for An Epistemological Foresight
Edited by Filippo Bignami, Niccolò Cuppini and Naomi C. Hanakata

Post-Utopian Spaces
Transforming and Re-Evaluating Urban Icons of Socialist Modernism
Edited by Valentin Mihaylov and Mikhail Ilchenko

Post-socialist Cities and the Urban Common Good
Transformations in Central and Eastern Europe
Maja Grabkowska

Post-socialist Cities and the Urban Common Good

Transformations in Central and Eastern Europe

Maja Grabkowska

Routledge
Taylor & Francis Group
LONDON AND NEW YORK

First published 2023
by Routledge
4 Park Square, Milton Park, Abingdon, Oxon OX14 4RN

and by Routledge
605 Third Avenue, New York, NY 10158

Routledge is an imprint of the Taylor & Francis Group, an informa business

British Library Cataloguing-in-Publication Data
A catalogue record for this book is available from the British Library

Library of Congress Cataloging-in-Publication Data
Names: Grabkowska, Maja, author.
Title: Post-socialist cities and the urban common good : transformations in Central and Eastern Europe / Maja Grabkowska.
Description: First Edition. | New York, NY : Routledge, 2023. | Series: Routledge contemporary perspectives on urban growth | Includes bibliographical references and index.
Identifiers: LCCN 2022025509 (print) | LCCN 2022025510 (ebook) | ISBN 9780367545734 (hardback) | ISBN 9781003089766 (paperback) | ISBN 9780367545741 (ebook)
Subjects: LCSH: Cities and towns--Europe, Eastern. | Cities and towns--Europe, Central. | Community development, Urban--Europe, Eastern, | Community development, Urban--Europe, Central. | Common good. | Post-communism--Europe, Eastern. | Post-communism--Europe, Central.
Classification: LCC HT145.E75 G73 2023 (print) | LCC HT145.E75 (ebook) | DDC 307.760947--dc23/eng/20220915
LC record available at https://lccn.loc.gov/2022025509
LC ebook record available at https://lccn.loc.gov/2022025510

ISBN: 978-0-367-54573-4 (hbk)
ISBN: 978-0-367-54574-1 (pbk)
ISBN: 978-1-003-08976-6 (ebk)

DOI: 10.4324/9781003089766

Typeset in Times New Roman
by SPi Technologies India Pvt Ltd (Straive)

I dedicate this work to my daughter, Matylda, who personally accompanied me in it from the very beginning.

Contents

Figures

Tables

Acknowledgements

This book concludes a research project titled 'Sharing Urban Space. Conceptions of the Common Good in Polish Cities after Socialism' funded by the National Science Centre, Poland, under grant agreement no. 2014/15/D/HS4/00750. I am very grateful for this opportunity. In terms of motivation and expert advice, I am indebted to my colleagues at the Department of Economic Geography at the University of Gdańsk and to my persevering proofreaders. My deep thanks go to the interviewees in Gdańsk, Kraków, and Łódź for sharing their time and insightful perspectives. Yet, my most heartfelt appreciation is for my family and friends, especially Piotr—without their unwavering support, this process would never come to an end.

Introduction

This book covers and dissects the transformations of urban common good in Central and Eastern European (CEE) cities in the post-Soviet Bloc era, but the issues I discuss fit into a larger framework of urban commonality. When I embarked on the process of writing in 2019, I already had the idea that the opening paragraph would refer to a bed-time story featured in a novel by Ethiopian-born writer Dinaw Mengestu. Told by a father to his son, the story is about a city which existed only 'as long as one person dreamed of it each night' (2016: 175). For many years, citizens in their sleep had duly engaged in conjuring up streets, parks, houses, and compounds to preserve them in the waking world, until one day they began to find it tiresome and boring. One by one, people would choose to cede this responsibility to others and dream about something else. In the end, there was only a single person left vowing to dream of the whole city and assure its continuance on behalf of all the citizens. Instead, he used his position to take control of the city and alter it according to his own vision. I thought of this passage as an adequate and universal metaphor of the notion of urban common good, intending to precede it with a sentence which would read: 'Cities make no sense without citizens being there to share in the experience of living within them'.

Little did I know back then that the global COVID-19 pandemic would occur in March 2020, followed by an escalation of the Russo-Ukrainian war almost two years later, both events reasserting the relevance of the topic I take up in this monograph. One of the manifold consequences of the former event is that it abruptly redefined the experience of shared urban living. With the spread of the virus across continents, more and more cities went under lockdown and their inhabitants' interactions became limited by restrictions preventing the transmission of the disease. Deserted public spaces and masked, socially distancing individuals became familiar images even in places as densely populated as Venice or Shanghai. More recently, these images were replaced by a yet more compelling visual documentation from bombed Ukrainian cities, as well as from towns across the border, which mobilised to offer refuge to the evacuating civilians.

The notion of 'common good' may be understood in many ways. Its interpretation as a general value or principle organising the relationship between individual and community interests goes back to Aristotelian thought. Such

DOI: 10.4324/9781003089766-1

conceptualisation is highly reliant on the local context—'the common good in an American inner-city neighbourhood would look vastly different from that in a farming community in Tibet because of different economies, cultures, citizens, problems, and opportunities' (Smith 1999: 633). However, context-dependency also applies to the substantive meaning of the common good, meaning a resource which is shared between users. According to Garrett Hardin (1968), sharing common goods in the long term leads to their devastation. Using the example of an overcrowded pastureland, he demonstrated his point that the future of the commons which are not privatised or taken over by the state is inevitably 'tragic'. This stance was questioned by Elinor and Vincent Ostroms (1977). Deriving from Paul Samuelson's (1954) distinction between private and public goods, they developed a category of common-pool resources to denote goods which are especially susceptible to depletion, being at the same time rivalrous and non-excludable. The former feature accounts for the fact that their utilisation by one user prevents or limits simultaneous use by other users. The latter consists of the inability to exclude users' access to the good. The Ostroms' perspective on the future of commons was more optimistic than Hardin's—they challenged his dilemma proving the effectiveness of collective governance of common-pool resources, provided that several conditions are fulfilled by the commoners (1990). These, inter alia, include setting and pursuing rules safeguarding the common interest, which brings us back to the procedural idea of common good.

Urban realms accommodate both approaches. On the one hand, the city—a socially produced and politically charged phenomenon—is founded on the principle of the common good. On the other, as a collectively governed and used entity, it may be also considered a commons, just like its components—urban space and infrastructure—meet the conditions for being recognised as common goods. In either of the two instances, the urban common good is subject to constant renegotiations. Contemporary cities cope with multiple internal and external tensions and challenges affecting the urban common good—deepening social inequalities, intensifying global migratory movements and climate change, to name only a few. Although a growing collection of studies undertaken by human geographers and scholars representing related disciplines touch upon this relation, there has been little research which would spotlight post-socialist cities of CEE (Sagan 2017).

To address this gap, the main purpose of this book is to investigate how the notion of urban common good transformed during the systemic transition from state socialism to a globalised market-based economy in CEE cities at the turn of the 21st century. Following World War II, the ideologically loaded notion of the common good legitimised intense urbanisation processes in Soviet Bloc countries across the region. Factories, housing developments, recreational spaces, nurseries, and other urban amenities designed to foster communal urban living were to 'belong' to urban citizens. This ideal was soon compromised due to systemic shortcomings and corruption. With the advent of a new socio-economic and political era around 1989, under extreme pressures of marketisation and privatisation processes, the communal way of thinking about the city was almost entirely rejected and just about vanished from public discourse. The recent decade saw a certain renaissance of the concept, mostly as a counter-reaction to the neoliberal urban agenda, which turned out to have a detrimental effect on urban commonality comparable to that experienced during the state socialist era.

Regardless of being increasingly acknowledged in cities of CEE, the urban common good tends to serve as an umbrella term, having different—often obscure—meanings and uses. Therefore, the research presented in this book reveals how the ideas of urban common good have been (re)constructed and put into practice in this part of the world. My research seeks to show the impact of the systemic transformation and accompanying processes on the perception of the common good in relation to urban spaces, and then establish patterns for the formation of the urban common good in post-socialist cities.

Being aware of the limitations of the latter denomination, further discussed in Chapter 1, I operationalise it by focusing on the commonalities which link the cities under investigation rather than on the differences which set them apart. In geographical terms, the research covers the countries which had gained independence from the Soviet Union following the Autumn of Nations in 1989 and which by the late 2000s became member states of the European Union. They were/are, in alphabetical order: Bulgaria, Czechoslovakia/Czech Republic and Slovakia, East Germany/Germany, Hungary, Poland, and Romania. However, to help us fully comprehend the specifics of relevant national and local contexts, more in-depth analyses concerning the currently evolving public discourse on the urban common good and varying conceptions of the notion among groups of urban stakeholders are focused on Polish cities. While the temporal scope of the research reaches back to the period of the post-World War II socialist state, the main study period ranges from 1989 to 2019. Quite unexpectedly, the timeframe was closed by the COVID-19 pandemic. While for the moment it represents a symbolic, if debatable, end of the 'old normal' (Batty 2020, Florida et al. 2021), it remains to be seen whether it becomes a caesura-bidding farewell to the optics of post-socialism in CEE.

A variety of research methods and techniques were applied in the course of the research to produce a comprehensive guide to understanding the relationship between the post-socialist city and the common good. The theoretical framework relies on academic literature reviews. Due to the transdisciplinary nature of the research problem, scholarly works from several areas of urban studies were consulted. Aside from references to the field of human geography, the research draws from urban planning and architecture (theories and concepts of planning and shaping urban spaces in relation to the urban common good), sociology and social psychology (changing discourses on the urban common good, perception of urban commonality), economy (the production and consumption of urban common goods), and political and legal sciences (linking the concepts of urban common good and governance). The empirical chapters feature a two-stage research procedure based on a qualitative mixed-method approach. It entails a critical discourse analysis of legal, print media and academic texts, as well as a multiple case study employing individual semi-structured interviews with representatives of key urban stakeholders in three Polish cities—Gdańsk, Kraków, and Łódź.

The first, literature-based part of the book consists of two chapters. **Chapter 1** presents an overview of theoretical approaches to urban commonality and develops a model of urban common good used as a frame of reference in subsequent chapters. Recognising the impact of neoliberalism on the recent processes of urban change in Europe, the clash of two resulting paradigms—of the city as a commodity and city as a commons—is discussed in detail. The last section signals specifics of the outlook on urban common good in CEE. This issue is further elaborated in **Chapter 2** which also explores the changing orientations of the urban paradigms during socialism and after. Consecutive sections are organised around three of them, namely, the city as a communal infrastructure, the city as a commodity and the city as a commons. They provide a comprehensive background for the empirical contributions in Part II, investigating the mechanisms behind the reinvention of the urban common good in Poland. **Chapter 3** is devoted to the analysis of conditions shaping the changing understanding of urban common good in legal, print media, and academic discourses, while **Chapter 4** offers a comparative perspective by different stakeholders in selected Polish cities via a multiple case study. Summary and concluding remarks are included in **Chapter 5**.

References

Batty M (2020) The Coronavirus crisis: What will the post-pandemic city look like? *Environment and Planning B*, 47(4), 547–552. https://doi.org/10.1177/2399808320926912

Florida R, Rodríguez-Pose A, Storper M (2021) Cities in a post-COVID world. *Urban Studies*, https://doi.org/10.1177/00420980211018072

Hardin G (1968) The tragedy of the commons: the population problem has no technical solution, it requires a fundamental extension in morality. *Science*, 162, 1243–1248.

Mengestu D (2016) All Our Names. In: Wakatama Allfrey E (ed.), *Africa 39: New Writing from Africa South of the Sahara*. London: Bloomsbury Paperbacks, 175–180.

Ostrom E (1990) *Governing the Commons: The Evolution of Institutions for Collective Action*. Cambridge: Cambridge University Press.

Ostrom V, Ostrom E (1977) Public Goods and Public choices. In: Savas ES (ed.), *Alternatives for Delivering Public Services: Toward Improved Performance*. Boulder: Westview Press, 7–49.

Sagan I (2017) *Miasto: Nowa kwestia i nowa polityka*. Warszawa: Wydawnictwo Scholar.

Samuelson P (1954) The Pure Theory of Public Expenditure. *Review of Economics and Statistics*, 36, 387–389.

Smith TW (1999) Aristotle on the Conditions for and Limits of the Common Good. *American Political Science Review*, 93(3), 625–636. https://doi.org/10.2307/258

Part I

Urban common good before and after 1989 in theory and practice

1 The city and the common good

In search of a common ground

Commonality in the city

From ancient times to the present day, the concept of city has entailed the paradox of unrelated individuals living together. I began working on this book with this particular thought in mind. Zygmunt Bauman (2003: 106) expressed it even better, claiming that '[w]hatever happens to cities in their history, and however drastically their spatial structure, look and style may change over the years or centuries, one feature remains constant: cities are spaces where strangers stay and move in close proximity to each other'. This quite apparent facet of urban commonality implies, among other conditions, the inevitability of sharing. Sharing *of*—public space, communal infrastructure, common challenges—but also sharing *between*—the younger and the older, the more and the less privileged, 'us' and 'them'. Such distinctions already breed a plethora of potential socio-spatial conflicts undermining urban commonality.

Sharing also comprises the opposition between 'division and distribution' and 'having or using in common' (Wiktionary 2022). Thus, one way to look upon the city is to see it as a collective resource, publicly produced and, as such, publicly maintained and sustained (Healey 2002: 1790). This perspective calls on Henri Lefebvre's (1974) premise of the social production of space, based on the dialectical relationships between spatial practices (*le perçu*) and representations of space (*le conçu*), which combine into spaces of representation (*le vécu*). A related trialectics of spatiality introduced by Edward Soja (1989) reiterates Lefebvre's concept by accentuating the role of the 'real-and-imagined' thirdspace in breaking out of the 'real' and 'imagined' limitations of firstspace and secondspace. It is therefore important to consider urban commonality within these three dimensions—a proviso which I will return to in the subsequent section on urban common good.

The unfading trend of urbanisation across the globe would suggest that the requirements of sharing the city with unfamiliar others do not deter billions of people from the urban experience. Nor does the resultant imperative of give-and-take. Any city dweller sooner or later comes to terms with the fact that the mediaeval *Stadtluft macht frei* principle only tells half the story. While offering freedom, the city curtails it at the same time. Relieving citizens

DOI: 10.4324/9781003089766-3

from certain norms, rules, and obligations, it compels them to adhere to others in return. Urban liberties and access to urban resources come at a cost, be it local taxes or conforming to local regulations. Urban commonality thus hinges on responsibility and commitment—both individual and collective, held by all urban stakeholders, although to varying degrees.

Throughout this book, I employ the term 'citizens' to denote a broader category of urban actors, reaching beyond inhabitants entitled to participate in the local decision-making, for example, through the institutions of representative democracy. I also include groups who may be officially deprived of this right, such as immigrants, temporary residents, or homeless people, as well as city users living outside the administrative boundaries of urban units, who nevertheless locate the centre of their vital interests within them. Such understanding of the term remains close to another notion of Lefebvre (1996), namely that of citadin—an urban dweller and user of multiple services, who holds prerogative powers towards the inhabited space, conceptualised as the right to the city. While Lefebvre refers to inhabitation, Mark Purcell (2002: 1936) cautions against interpreting it as an exclusionary criterion:

> The essence of Lefebvre's vision is to favour those who inhabit space over those who own it. It is not to favour local inhabitants over inhabitants at wider scales or, I would argue, to favour urban inhabitants over rural ones. The right to the city, in particular, leads easily to a privileging of the local urban scale and thus into the local trap. In using Lefebvre's ideas to help us to imagine new forms of democracy, it is critical not to privilege the urban—as a scale—a priori.

Similarly, Rob Shields (2013: 347) is cognizant of the role of outsiders in the process of city-making, regarding the openness of the urban as a key pathway to innovative ideas and vehicle for creative social-economic development.[1]

An inclusive definition of citizenship invokes the question of urban community, whose connotations usually involve a sense of shared identity, as in Aristotle's idea of a self-contained polis, in which citizens holding identical values work together to achieve a common goal. Per contra, populations of contemporary cities form a motley of communities that are hardly likely to unite solely on the assumption of a common background or ideals. Hence, Ash Amin and Phillip Howell (2016: 10–11) propose a definition of community based on an alternative etymology of the word. Instead of understanding it as 'coming together as one' (*com-unus*), they argue for acknowledging another meaning, derived from a common obligation (*com-munis*). While the authors make this distinction in relation to the commons, shifting the emphasis from an abstract motivation (why anything should be done together) towards a specific purpose and collective action (what needs to be done together), their recommendation seems also relevant for an exploration of urban commonality.

As a matter of course, commonality in a city also grows out of the tug of war between self-centredness and altruism, fuelling the opposing forces of individualism and collectivism. As the balance tilts back and forth according to exigencies of Zeitgeist, it leaves noticeable traces in the urban palimpsest (Montgomery 2013). The conflicting demands of the collective and the individual are in fact two sides of the same coin. Collectivism allows individuals to achieve otherwise unattainable goals at the expense of independence and de-prioritisation of their own interests. Individualism forgoes the reckoning with others for the sake of one's autonomy, but it leaves one single-handed. This dualism remains in force whether we sympathise with John Donne's belief that no one is a solitary island or share the viewpoint of Margaret Thatcher, who downplayed the significance of society against the individuals' self-sufficiency. Cities across the globe have developed not despite of but owing to the interplay between the two perspectives.

Commonality in the city therefore relies on the combination of practices of sharing and the connoted responsibility, as well as on the dialectics of individuality and collectivity. It is not given once and for all, but constantly negotiated, and it depends on a variety of equally changeable circumstances. The underpinning of this transience and incoherence is well illustrated by the concept of throwntogetherness, proposed by Doreen Massey. Considering space as 'an event', or 'an ever-shifting constellation of trajectories' (2005: 151), it explores how incidental encounters of human and non-human actors coming together at particular moments in time generate conflicts through which the urban space is produced. While such conflicts follow from collisions of individual and collective interests of different urban stakeholders, these interests are conceived and pursued via strategies of competition and cooperation. They account for the political aspect of commonality, that is, the common good.

What makes the urban common good?

Just like the root of the word 'community', the heading of this section also has a double meaning. It can be read either as an invitation to muse on the ethics of urban commonality or to discern the constituents of urban common good. The former relates to moral standards that regulate social behaviour incorporated in urban commonality. The latter deals with the identification of elements of common good when the concept is transferred to the local level. However, the two issues are interconnected. Together, they shape the ideal that urban societies strive for when facing the multiple challenges of the contemporary world: climate change, social and spatial injustices, global migration crises, unequal distribution of resources, dysfunctions of democracy, pandemic risks, environmental pollution threats, and many, many more.

In order to investigate the 'goodness' of urban commonality, I turn once again to Ash Amin. Coming up with a touchstone of a 'good city', he understands it as 'the kind of urban order that might enhance the human experience' (Amin 2006: 1009). While the specifics of this order largely depend on the particular historical timeframe and socio-cultural context, the author seeks universal qualities which could provide the basis for a good urban life under various circumstances. His prototype revolves around four measures of urban solidarity: repair, relatedness, rights, and re-enchantment.[2] In brief, repair guarantees non-exclusionary urban resilience. Relatedness builds on social justice to result in 'an equal duty of care towards the insider and the outsider, the temporary and the permanent resident' (1015). The third R denotes the universe of rights to the city postulated by Lefebvre, as well as securing space for pluralism and dissent in the urban public sphere. Re-enchantment involves reclaiming the lost pleasure from urban commonality, and, in particular, concerns association and sociality performed in all types of public spaces and through public art.

Amin's attempt, which he himself dubs 'practical urban utopianism', may be considered as a valuable input in the discussion on the indispensable features of a good, that is, virtuous, city, one which embraces and supports the needs of its citizens, regardless of their background, power position, and other standings. Its solidarity-driven approach provides a robust platform for the development of urban common good. Such a proposition, however, does not rule out conflict. The latter is a natural condition accompanying any situation of commonality. What counts is the way it is managed or accommodated—whether via an antagonistic struggle of enemies for life or by means of pluralistic agonism providing space for reciprocity and mutual respect (Mouffe 2000).

This brings us to the second issue signalled in the section heading. First of all, it should be underlined that the questions of what constitutes the common good and how to achieve it have been debated for centuries. Premodern conceptions were centred on the archetype of political community, overshadowing the role of individuals. Aristotle imagined the common good as the highest priority, guiding citizens of the polis regardless of the form of government in force. He expected it to be put before individual interests in the name of the welfare and happiness of the community as a whole. A similar supposition may be found in the early Catholic philosophy represented by Thomas Aquinas, even though, convinced of people's propensity for following their own concerns, he sought assistance from the government to keep the citizens on the right path.

The modern shift of emphasis to the individual, which occurred under the auspices of liberalism, disrupted the presumption that the common good should be collectively sought by and benefit the community first and foremost. It would now be regarded as a sum of rather uncoordinated contributions to the common weal, following from individual purposes and motivations, and regulated only externally. The metaphor of the 'invisible hand' provides one example of a potential regulatory mechanism; another would be John Rawls's (1971) confidence in the rationality of individuals automatically translating into collective rationality (Koczanowicz 2021). So far, both explanations have proved inefficient, undermining the integrity of the liberal-individualist approach to the common good.

Meanwhile, the community-oriented, or communitarian, vision of the common good was not abandoned in modern times. Initially marginalised and only continued in the more or less utopian works of sociologists and philosophers, it culminated in the social projects of the 19th and, especially, the 20th century. Its career in Central and Eastern Europe will be discussed in Chapter 2; here, my main focus lies on Western Europe. This is where, since around the 1980s, the communitarian approach re-emerged in the works of academics who opposed the eclipse of the community caused by disproportionate individualism (Sandel 1982, Taylor 1989). Since then, it has successively gained ground in the public sphere, partly due to the growing disillusionment with the neoliberal model. However, the communitarian approach is not free from deficiencies either, one of them being the idealisation and overemphasis of the unifying force of the community's shared values and identity in reaching the common good. As much as the significance of community per se should not be underestimated, one of the risks entailed in this type of logic is the conceivably dangerous liaison with the idea of national community and nationalism.

Thus, according to Leszek Koczanowicz (2021), neither of the alternative standpoints—liberal nor communitarian—offers a more universal recipe for the common good in a pluralist democracy. Drawing on the concepts proposed by John Ryder (2013), inspired by the thought of John Dewey, he goes on to propose a pragmatist 'third way' in conceptualising the common good. It is based on two principles. The first consists of anti-ideology, understood as strategic resistance to ideological influence and authority. Helping to avoid the trap of uncritical valorisation or prejudice, it facilitates unbiased searching for effective here-and-now solutions to problems as they arise. The second principle calls for looking beyond the interests of a single individual or community by acknowledging the stakes of other individuals and communities in all their multiplicity and variety. This cosmopolitan perspective highlights the advantages of mutual interaction towards common ends, while at the same time allowing for the possibility of conflict in the process, in line with the notion of agonistic pluralism. On the whole, this programme also corresponds with the *com-munis* take on community and the four R's of the 'good city'.

Despite being a contested concept, whose intricacy does not disappear when the discussion switches to the urban scale, the common good amounts to an essential urban attribute. As maintained by John Friedmann (2000), it is the common good that furnishes the city with a sense of local identity and a political community. Deprived of these two features, urban commonality becomes futile, and the city itself turns into an administered entity which 'might as well be a hotel managed by some multinational concern' (2000: 465). Therefore, when looking for a conceptual scaffolding upon which to build a model of urban common good, we should not only acknowledge the weight and force of socio-spatial affinities and their interplay, but indeed make them central.

This view is embraced in the conceptual framework put forward by Robert Asen (2017: 331), who formulates the common good as a relational practice 'of articulating mutual standing and connection, recognizing that people can solve problems and achieve goals—and struggle for justice—through coordinated action'. While the author distances himself from the term 'common good', consistently referring to 'public' good instead, his definition is fully applicable to the former.[3] In Asen's interpretation, a 'dynamic and mobile' public (common) good 'circulates in society, connecting people's perceptions and actions to their relationships with each other and the worlds they inhabit' (332). Emphasising relationality, the approach clearly focuses on the links and interactions within the public sphere. Accordingly, the common good is presented as a connecter binding the three elements, rather than a vessel which simply holds them together through containment.

In general, Asen's model meets the characteristics of urban commonality, such as multiplicity, community as responsibility, and throwntogetherness. Adapting it to the urban scale results in the following reformulation: urban common good circulates in urban society, linking collective human actors ('people's relationships with each other'), dominant rationales ('people's perceptions and actions') and images of the city as a collective resource, or urban paradigms ('worlds people inhabit'). The first two elements are almost self-explanatory. Collective urban actors stand for stakeholders in the city's commonality. Their relations may vary from loose, when they tend to act individually, to tight, when they form close-knit communities. The dominant rationales explain the actors' behaviour in the urban public sphere—they underpin perceptions and legitimise actions. As for the third ingredient of urban common good, it refers to Soja's thirdspace. It is based on the reconciliation of the material urban world with its imagined representation, which jointly constitute the image of the city in terms of its commonality.

The model's inherent relationality underlines the two-way nature of the relationships at work. For instance, collective urban actors both shape the dominant rationales and are shaped by them. A particular urban paradigm holds because of the specific perceptions and actions performed by the actors maintaining it, but it may be overthrown by the same actors once their rationales change substantially. Hence, the communication between the three components relies on discursive social practices. According to Norman Fairclough (2001: 163), discourse plays a part in the inception, development, and consolidation of social change. It is in the public discourse that the first hints of epistemic undercurrents emerge. Provided that they do not sink, but surface and gain strength, they may become able to modify the mainstream regime of the public sphere.

Depending on the discursive orientation of the three components towards the making of the common good, they can come together or diverge. Having observed how the neoliberal public sphere, strictly reliant on individualism, undermines the common good, Asen comes up with an alternative—a multiple networked public sphere, offering conditions conducive to coordinated collective action which works in favour of the common good. To investigate these dynamics more closely, another dimension needs to be added to the model: one that would include the effects of the interplay of the dialectic

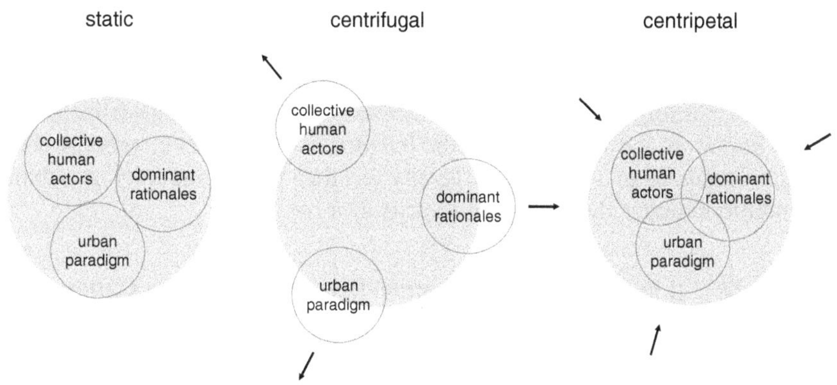

Figure 1.1 Possible dynamics in the model of urban common good.

forces of individualism and collectiveness. Taking into account their oppositional character, three possibilities arise (Figure 1.1). When the forces are equal, they counterbalance each other and the common good remains in relative equilibrium. When either of them gains advantage, the elements of the model gravitate outward (advantage of the centrifugal force) or inward (advantage of the centripetal force), affecting the relationships involved in the making of urban common good.

A comprehensive conceptualisation of centrifugal and centripetal dynamics was first applied in human geography by Charles Colby (1933), who studied their impact on the functional organisation of urban areas. According to his findings, centrifugal forces contribute to the dispersion of certain urban functions, pulling them out of the centre towards the relatively more attractive peripheries. Correspondingly, centripetal forces translate into processes of agglomeration by holding other functions within the centre, whose attractiveness grows as a result and prompts further gravitation in this direction. Economist Paul Krugman (1994: 243) has explored in more detail the tension between the two energies: the centripetal, which 'tend[s] to pull population and production into agglomerations' and the centrifugal, which 'tend[s] to break such agglomerations up'. His examples of the former include central locations, access to markets and products, as well as knowledge spillovers, whereas the latter comprise, inter alia, urban land rent, congestion, and pollution. In political geography, centrifugal and centripetal factors, respectively, hinder and favour the assembly of regions within a country into one effective unit (Hartshorne 1950). The centrifugal tendencies resulting from the regions' separate and divergent interests are counterbalanced by the pull of such binding forces as 'the idea of the state, a purpose or set of purposes' (129).

Similar dispositions of the centrifugal–centripetal dyad may be observed in the interplay of collectivism and individualism in the advent of neoliberalism, which considerably transformed the approaches to urban common good. By the end of the 20th century, the neoliberal turn overlapped with the rise of individualism; the resulting precedence of autonomy, self-fulfilment, and private property over communal values created a tension that worked in favour of the centrifugal forces. As a result, the elements of the neoliberal public sphere—atomistic individuals following the rationale of privatism within the paradigm of city as a commodity—were scattered away from the urban common good. To retrieve the balance, in a sense as a response, the collectivist stance of the right to the city reclaims the 'community of urban communities' under the banner of the city as a commons.

The neoliberal imprint: city as a commodity versus city as a commons

The constant renegotiations of the urban common good occur in the public sphere, which is not a fixed concept either. Malcolm Miles (2009: 133) recommends thinking of it as an ephemeral process, spanning divides between public and private realms and subject to change over time. Thus, to fully understand the ongoing articulations of urban common good, it seems necessary to look back to the origins of the public sphere and trace the trajectory of changes that brought us to where we are now, in the early 21st century. The discussion in this section is still limited to Western Europe; therefore, the retrospect takes us three centuries back to the emergence of the bourgeois public sphere identified by Jürgen Habermas (1989). It will be used as a background for the early capitalist model of urban common good. I will then describe its transformation, via Nancy Fraser's (1996) proposition of post-bourgeois counterpublics, into the neoliberal public sphere as described by Asen (2017). Next, drawing on Asen's ideas, I will turn to a discussion of the most recent configuration of urban common good in a networked public sphere.

Thus specified types of the public sphere serve as contextual backdrops for the changing constituents and dynamics of urban common good. Nonetheless, the consecutive order in which they are presented should not in any way suggest a linear evolution. Rather, it reflects the succession of ways in which complex and multiple modes of publicity affect the urban common good. As they tend to transform over time, they may also overlap, as indicated in the second column of Table 1.1.

The modern public sphere derives from the city. This was where it came into being and from where it expanded. In Western Europe, back in the 18th century, the scale and density of urban environments, which shortened the distance in social interactions and increased their intensity, together with the unparalleled socio-economic conditions of the epoch, were conducive to the emergence and consolidation of the bourgeois public sphere. According to Habermas (1989: 27), people's coming together as a public was possible due to the separation of the state and society in the post-feudal system. From now on, matters of public interest and the common good were not to be imposed by the church or secular authorities, but decided together, through rational reflection and consensus-building, by stakeholders with equal communicative power, who cultivated a cosmopolitan consciousness.

Table 1.1 Transformations of the model of urban common good in Western Europe

Type of public sphere	*Approximate timeline*	*Elements of urban common good*			*Dynamics of urban common good*
		collective human actors	*dominant rationales*	*urban paradigm*	
Bourgeois public sphere	18th to 19th century	the bourgeois public	cosmopolitan consciousness, universality	city as a space for debate	centripetal →static
Transformed bourgeois public sphere	19th to late 20th century	pseudo-public, mass society	manipulative publicity, welfare state	city as a space for consumption	centrifugal
Post-bourgeois counter-publics	19th to late 20th century	excluded groups, marginalised publics, urban social movements	social justice, recognition, urban question	city as a space for representation and/or contestation	centripetal impulses
Neoliberal public sphere	1980s to the present	atomistic individuals	individualism, competitiveness, privatism	city as a commodity	centrifugal
Networked public sphere	1990s to the present	urban movements, local communities, commoners,	right to the city, spatial justice	city as a commons	centripetal

Discussions and the forging of public opinion took place in literary salons and cafés, accounting for the physical nesting of the public sphere in the generally accessible urban public space. Although the bourgeois public sphere undoubtedly positioned the **city as a space for debate**, the questions of its inclusiveness and the equal status of participants have been subject to academic critique. Habermas envisioned the bourgeois public sphere as open to everyone, but in practice, the access was restricted to the privileged upper and middle classes. Apart from the lack of representativeness of the entire urban society, the universality of the common concern precluded any substantial differences of opinions or perspectives. Therefore, the flawed mechanism of public participation, in principle oriented towards the common good, inhibited the centripetal energy and brought the whole model to a static balance.

Moreover, once established, the public sphere soon underwent degeneration. Habermas attributes this to the merging of the private and public in what he terms the process of re-feudalisation of the public sphere. He found that due to processes of democratisation and increasing state intervention, the public was less inclined to actively engage in the common debate. Paradoxically, the institutionalisation of the public sphere thwarted effective communication among citizens and diverted their interests away from civic to private matters. Moreover, the welfare state replaced public engagement, and the dynamics of urban common good, previously brought to a halt, were redirected outwards. Within the transformed bourgeois public sphere, the pseudo-public of mass individuals symbolically abandon the politically imbued venues for the mushrooming department stores, to which they are successfully drawn by the advertising discourse of the mass media. The new predominant paradigm of the city renders it a **space for individual and collective consumption**—the former taking place through the market, the latter through the state and resolving to socialised goods and services, such as housing and public transport (Castells 1977).

Shortly after Habermas published his theory, it was put to a test. The events of the 1960s highlighted its neglect of non-bourgeois actors engaged in the public domain but dominated by the hegemonic majority. It was a time when public spaces across the world were reclaimed by dissenting publics. Protesters against the war in Vietnam took to the streets in US cities and Western European capitals. The year 1968 was marked by, inter alia, citizen and student demonstrations in Czechoslovakia and Poland during the Prague Spring and March events, the occupation of French universities followed by general strikes in May, and the Tlatelolco massacre of peacefully protesting civilians in one of the Mexico City squares. Around the world, cities became arenas of struggles for social and political change, and all sorts of social movements mobilising citizens in the urban milieus began to be increasingly recognised as important collective actors in this transformation. The urban question, raised by Manuel Castells (1977), publicised the accretion of local-based issues.

The Habermasian approach to the public sphere, criticised for its unitarity and exclusiveness, fails to encompass these phenomena. In contrast, the alternative concept put forward by Fraser endorses a multiplicity of publics and responds to social inequality, as well as calling for a revision of issues to be debated. Her notion of the postbourgeois counterpublics thus acknowledges the existence and performance of marginalised groups which counter the hegemony by forming subaltern counterpublics—'parallel discursive arenas where members of subordinated social groups invent and circulate counterdiscourses to formulate oppositional interpretations of their identities, interests, and needs' (1996: 123). As an example, Fraser mentions a feminist counterpublic that aims to overcome gender inequity in the mainstream public sphere by producing an array of its own means of expression and participatory practices. This also applies to other social movements actively participating in the reworking of the dominant bourgeois paradigm (cf. Negt and Kluge 1993). However, until the 1980s, their 'urban' designation was related to their territorial grounding rather than a concentration on strictly urban matters.

Fraser's argument is founded on the concept of social justice that is not limited to redistribution but extends to involve recognition. The postbourgeois counterpublics hence feature the **city as a space of representation and contestation**, seemingly rife with potential for collectivity and inclined towards collective action. Yet, the existing pluralism does not translate into systemic mobilisation for the sake of urban common good. Centripetal impulses arise and disappear, but the publics do not communicate with one another, lacking a common rationale that would directly address the *com-munis* aspect of urban community. The right to the city advocated by Lefebvre already in 1968 took hold only several decades later. Acting as a discursive adhesive, it effectively bound the elements of urban common good together; this process was catalysed by the crisis of neoliberal capitalism.

As indicated in the previous section, the neoliberal turn of the 1980s left a heavy imprint on urban environments. On the one hand, cities became overburdened with the tasks of the retreating state. On the other, the general dismantling of the public sector increasingly drove urban policies away from the 'welfare', ushering in the market logic of competition, efficiency, and profit maximisation. While Neal Brenner and Nik Theodore (2002) draw attention to the many guises of the neoliberal project and its context embeddedness by promoting the term 'actually existing neoliberalism', they also specify North American and Western European cities as 'strategically crucial arenas in which neoliberal forms of creative destruction have been unfolding' (20). The notion of creative destruction refers to the concurrent disruption of the hitherto existing institutional arrangements and to their readjustment, resulting in such processes as privatisation, enclosure, deregulation, and shift of responsibility from the collective to the individual.

For the last four decades, as a result of these reconfigurations, the neoliberal order has been articulated in the built environment through the implementation of large-scale megaprojects, emergence of gated communities, and commercialisation of public spaces. However, the emanant urban paradigm no longer pertains only to the spatial dimension. Instead of being seen as 'a space for' (debate, consumption, representation and/or contestation), the city comes to be perceived as an entity in itself. In this particular case—**a commodity**, possessing both use- and exchange-value and therefore metamorphosing from 'a place of business' to 'a business' (Carlos et al. 2017). In analogy to David Harvey's (1989) observation of the changing mechanisms of urban governance, we may see this as the transition from managing the city's resources (managerialism) to treating the whole city as an enterprise (entrepreneurialism).

Although the commodification of urban life had been progressing throughout the 20th century, it reached an unprecedented scale in the era of neoliberalism. This is when the collective decisions of citizens 'recast as consumers' are 'transformed into questions of individual need and choice' (Fenton 2018: 30).

Hence, Asen's likening of the neoliberal public sphere—dominated by privileged atomistic individuals following private interests and refraining from acknowledging difference—to the likewise unitary and exclusionary model of the bourgeois public sphere. The model's centrifugal disposition disaggregates rather than combines for urban common good. Recognising competitiveness, underpinned by individualism and privatism, as the key mode of social interactions under neoliberalism, Asen notes its detrimental effect on citizenship and public engagement. Combined with the corollaries of entrepreneurial governance, such as insufficient public transparency and empowerment of the private agenda, the overall outcome inevitably contributed to the erosion of democracy and, implicitly, urban commonality (Graeber 2013, Mayer 2018).

It is commonly agreed that the neoliberal programme culminated in the global financial crisis of the late 2000s. Although neoliberalism persisted, its hegemony has become increasingly contested (Peck et al. 2012). Urban citizens, particularly affected by the crisis and the subsequent austerity measures, repeatedly rose up to protest. The revival of public participation and political engagement, coupled with the use of information and communication technologies, facilitated worldwide social mobilisation. Just as traditional print media paved the way for the 18th-century bourgeois public sphere, social media platforms enabled the symbolic if non-uniform '99%' to come together and effectively publicise the common message, calling against the ills of the neoliberal city and for spatial justice.

Although the crowds taking over urban squares across the world—in Cairo, Madrid, New York, London, and Rome—have become emblematic of citizen uprisings, it is impossible to ignore their volatility and dispersedness. In this respect, Margit Mayer's (1999: 209) portrayal of the urban movements of the late 20th century as 'heterogenous, fragmented, and even polarized' remains relevant and should be taken under consideration. However, unlike in previous decades, the variety of actors and diversity of interests did not preclude a scenario of a politically charged, networked response. The Indignados and Occupy movements that grew strong in 2011, as well as their manifold local followers, did not just turn cities into temporary arenas of protest, but gave rise to a long-term reassessment of the urban paradigm. As a consequence, apart from being questioned, the neoliberal standard of city as a commodity was eventually countered with a rival proposition—that of **city as a commons**.

These developments were both accompanied by and contributed to a critical reworking of publicness. Its yet another incarnation took the form of what Asen terms a networked public sphere—a rendition of Fraser's subaltern counterpublics, embracing the idea of multiplicity, but enhanced by new possibilities for coordinated action. He describes this concept using the example of concerted bottom-up efforts undertaken to resist a neoliberal reform inhibiting local access to public education. Joint engagement and cooperation between usually disparate actors operating within several communities, supported by mutual recognition and effective communication, yielded success. Asen attributes this to the power of their working together over specific concerns, surmounting the lack of uniformity or homogeneity:

> In connecting school and community, they cast education not only as a means of enhancing one's individual competitiveness, but as a way of strengthening bonds while also enabling individuals to realize their potential.
>
> (343)

Asen's acknowledgement of interconnectedness confirms the intuition of Natalie Fenton and John Downey (2003), who had anticipated two conditions for the possibility of redressing democracy through a counterpublic. The first consists in the quality of the rationale behind participation, that is, whether or not it surpasses 'enlightened self-interest'. The second one entails the counterpublic's ability to renounce the fortified position of the dominant neoliberal approach. Within the urban dimension, both conditions are met in the discourse and practices claiming the right to the city. It is in the same manner that David Harvey (2008, 2012), drawing on Lefebvre's *le droit à la ville* (1968), envisages a crucial transgression of urban citizenship:

> The right to the city is far more than the individual liberty to access urban resources: it is a right to change ourselves by changing the city. It is, moreover, a common rather than an individual right since this transformation inevitably depends upon the exercise of a collective power to reshape the processes of urbanization.
>
> (Harvey 2008: 23)

The considerable capaciousness of the term, relating both to non-exclusionary entitlement to opportunities provided by the city and legitimacy of active participation in collective decision-making, accounts for a myriad of reinterpretations (Attoh 2011). Nonetheless, the right to the city remains a potent slogan, capable of bringing together the 'community of urban communities', which become connected regardless of their fragmentation (Blokland et al. 2015). This partly explains its popularity in urban research across the world (Fahmi 2009, Friendly 2013, Samara et al. 2013, Wong and Liu 2017). Other factors contributing to its global career include its compatibility with the *modus operandi* of the new media, alignment with glocalisation tendencies, and the universality of contemporary social and environmental challenges 'pushing into action diverse groups of residents, including middle class and intelligentsia, whose voice resonates in the public debate' (Domaradzka 2018: 610).

Out of the polyphony comes a fairly unison communiqué, taking a stand on, among other issues, the commodification of housing and residential segregation (Barbero 2015), appropriation of public space and urban green open areas (Bodnar 2015, Carmona 2015), neoliberal restructuring of public transport (Enright 2013), gentrification and touristification (Lees et al. 2015, Colomb and Novy 2017). Voices of opposition and protest do not stop short of concrete propositions for reclaiming the right to the city. For instance, editors of publications based on World Social Forum debates identify the following six solutions, hailing them 'alternatives to the city as a commodity': participatory budgeting, systemic measures against forced evictions, housing and employment co-operatives, collective and communal forms of land tenure, complementary and local currencies, urban and peri-urban agriculture arrangements (Cabannes and Delgado 2015).

In academic discourse, the right to the city teams up with complementary rationales of 'the just city' (Fainstein 2010), 'spatial justice' (Soja 2010), 'cities for people, not for profit' (Brenner et al. 2011), and 'the politics of encounter' (Merrifield 2013). Together, they fit into a resurgent narrative of urban commonality and provide a platform for a reappraisal of the previously abandoned idea of urban common good. In the same vein, the right to the city serves as an underpinning for Sheila Foster and Christian Iaione's (2016) concept of the city as a commons. Their understanding of the city as 'a shared resource that belongs to all of its inhabitants' (288), touches upon one of the dilemmas of urban commonality—the tension between concurrent openness and potential for rivalry, which entails the risks of appropriation and exclusion:

> On the one hand, it is the openness of cities that allows them to make and remake themselves, and to compete for the people and goods that help to grow and sustain them. On other hand, the city has finite resources that, by virtue of being open, are subject to congestion and exhaustion, rendering those cities vulnerable to rivalry. This rivalry often puts in conflict different kinds of commons, or commons claims, leading not only to the sacrifice of one urban good (e.g., a park or garden) for another (e.g., housing) but the sacrifice of the needs of the socially and economically powerless for the desires of the more powerful.
>
> (Foster and Iaione 2016: 334)

The main challenge for the citizens cast in the role of commoners is thus to remain in control of the city as a commons, while also managing it in a just and sustainable way, applying lessons learnt from Elinor Ostrom (1990). This implies the empowerment of citizens through their involvement in a specific model of governance. Foster and Iaione conceive it as a collaborative and polycentric system of city-making, in which power and influence are reshuffled away from the centre and towards an engaged public (335). As such, it remains consistent with Asen's centripetal vision of common good in a networked public.

The programme put forward by Foster and Iaione may seem radical, but the commons framework has been campaigned for also by other authors, who see it as an alternative to the paradigm of city as a commodity. Michael Hardt and Anthony Negri (2009: 250) prefer the singular form of the noun. Their 'common' stands for 'people living together, sharing resources, communicating, exchanging goods and ideas, all of which is produced in the 'factory' of the contemporary city. Peter Linebaugh (2008) recommends the use of a verb—commoning—which denotes activity. Building on this approach, J. K. Gibson-Graham et al. (2016: 195) adopt a definition of commoning as a relational process of 'negotiating access, use, benefit, care, and responsibility'. They provide a concrete empirical example of how, through this process, citizens of Newcastle mobilised to common the city's polluted atmosphere. Finally, David Bollier (2014: 26) derives a formula of commons which combines 'a distinct community with a set of social practices, values and norms that are used to manage a resource'. What connects all these perspectives is that they shift the emphasis from the substantive aspect of the commons (shared assets or resources) to the procedural (social protocols and governing practices). An important aspect of this approach is its constructive potential. While the recent reappraisal of urban commonality originated out of opposition to the neoliberal status quo, the commons paradigm accommodates moving from protest to action. For instance, claiming the right to affordable housing may be successfully paired with arrangements of systemic support for co-housing initiatives.

Summing up, what follows from this brief chronological overview is that the parameters of urban common good are highly susceptible to the structural changes of the public sphere. Its transformations over the last three centuries have fostered the development of the presently reigning paradigm of city as a commodity and its ascending alternative of city as a commons. The neoliberal turn thus comes out as a triggering factor for urban commoning as a coping strategy, but also a programme which brings the components of the model of urban common good together, closer than they have ever been in the history of Western European urbanity. Due to the current instability brought about by the COVID-19 pandemic and the already felt repercussions of the quasi-global crisis following Russia's attack on Ukraine, we do not know what even the most immediate future holds, let alone what to expect in terms of the seemingly imminent transformations of the public sphere and the urban paradigm. I shall thus leave the analysis at this quite compelling moment in time to turn to the last subsection, which finally introduces us to Central and Eastern Europe.

Geographies of urban common good

Bearing in mind the declaration from the Introduction that the book concerns the post-socialist city and urban common good, it is time to explain why the discussion so far has been mostly limited to Western Europe. There are two main reasons. Firstly, my aim was to provide a general background for the Central and Eastern European versions of the model of urban common good that unfold in the following chapters. Giving prominence to Western European cities is justified by the fact that in the concluding decades of the 20th century, societies of the former Eastern Bloc looked westward for inspiration and with aspiration. Explaining the benchmark position of the West for transitioning from socialism and the disillusionment that ensued,

Ivan Krastev and Stephen Holmes (2019) use the metaphor of a light which brought promises of liberal democracy and capitalism but faded shortly after. Secondly, many of the Western European and CEE cities once separated by the Iron Curtain have now integrated as members of the European Union. Meanwhile, although absent from the Community acquis, the notion of the common good has lately become acknowledged in the urban agenda of the EU—for instance, the New Leipzig Charter (2020) adopts it as a guiding principle for urban governance and planning.

There is no agreement on a particular threshold that would denote the emblematic beginning of the end of socialism in the CEE region. In chronological terms, the first domino stone fell in Poland on 4 June 1989, when democratic opposition gained victory in the first partly free parliamentary elections held in the Soviet bloc. Yet it is the fall of the Berlin Wall, which took place five months later, that is more often recognised as the tipping point (Petsinis 2010, Krakovský 2018).[4] Of the two milestones, the latter is perhaps more powerful in terms of symbolism, with images of hammer-equipped East and West Germans rushing to participate in the collapse of the detested Iron Curtain. While cities, in general, are vital arenas for mobilisation, civic participation, and solidarity, reconsolidation of a divided city makes an all the more illustrative case in point. Documentary photographs and footage of reunited Berliners crowding atop the severed concrete slabs have not lost their emotional impact over time. Reflecting on those images today, one may speculate whether the actors of this carnival-like celebration were indeed looking forward to an instant reconciliation of the East and West, confident that the elimination of the physical barrier would at once bring the two regions and perspectives together.

Similar expectations resound in the academic discourse of that time and partly explain why much of the research on European post-socialist cities have relied on transplanting Western concepts and theories. As demonstrated by Slavomíra Ferenčuhová (2012), urban researchers of early post-socialism either tended to evade theory-building or used existing concepts conforming to the paradigm of 'catching up with the West'. Some authors viewed CEE through the lens of 'delayed economic and urban modernization' (Enyedi 1996) or 'under-urbanization' (Szelényi 1996), and others regarded the transformation period as a transitory stage of returning to the 'natural path of evolution of the West-European cities' (Mulíček 2009: 159) or, put metaphorically, as a liminal passage in the process of maturing and growing up (Czepczyński 2008). Until today, there has been an ongoing debate on whether the European 'East' indeed catches up with the 'West' or rather follows a distinct path of its own (Pickles and Smith 1998, Hörschelmann and Stenning 2008, Pobłocki 2017). For instance, a group of scholars has gone as far as to argue that the extreme version of neoliberalism applied in CEE countries after 1989 in fact led to an outperformance of the Westernisation model (Kideckel 2002, Chelcea and Druţă 2016).

In an attempt to systematise this diversity, Judit Bodnár (2001: 14–22) distinguishes three approaches to the East-West divide in theoretical standpoints in urban studies. The first emphasises the distinctness of socialism as a departure from capitalism, the second reduces the difference between the two to 'merely quantitative' discrepancies, and the third assumes that the East has always been different from the West, both in quantitative and qualitative terms. If we imagine the East and the West as two threads, the first situation (Socialist versus Capitalist Urban Logic) accounts for their separation after previous intertwinement, the second (Unified Urban Logic) refers to a slight loosening of their twist, and according to the third (Historical Continuity of Idiosyncratic Eastern Features), they have never been woven together in the first place, but always ran parallel to each other. Presenting the three comparative dimensions as 'ideal-typical intellectual traps', Bodnar recommends exploring their combinations, and this is the stance I choose to adopt in this book. To illustrate the complicated nature of conceptual mismatches in studies of processes observed in European cities with and without the experience of state socialism, I will refer to commoning as an example. Not exactly an originally urban phenomenon, it seems relevant in the context of the contemporary paradigm of the city as a commons, especially that, as demonstrated in Chapter 2, the latter has been adopted also in CEE.

A look into the historical evolution of the traditional commons in Europe reveals that the incongruity between the East and West is deep-rooted, partly extending to pre-capitalist societies. West of the River Elbe, institutions of collective management and use of natural resources formalised already in the late Middle Ages (Laborda Pemán and De Moor 2013). In the eastern part of the continent, analogous processes occurred much later, due to the region's relatively slower pace of urbanisation and market development. The persistence of archaic socio-political institutions based on kinship, noble privilege, and serfdom translated into fewer market risks and environmental pressures for the commons to emerge and expand. Interestingly—particularly in the context of the current COVID-19 pandemic—Daren Acemoglu and James Robinson (2012) evidence the impact of the bubonic plague in mid-14th-century Europe on the divergent institutional development in the East and the West. Faced with population losses and the resulting labour shortages, both regions adopted entirely different coping strategies. The former drifted further towards extractive regimes, embracing the feudal exploitation of serfs, while the latter was rebuilt on the basis of inclusive institutions, increasing labour effectiveness, thanks to liberation of the peasantry. Around the 19th century, the enclosure movement and top-down privatisation in Western Europe dismantled the fully formed and resilient common-property regimes. Yet the gap between the two institutional systems continued to widen, eventually turning into a chasm after the Iron Curtain was pulled down following World War II (WWII).

Forced collectivisation under the CEE communist regimes to some extent resembled the enclosure and privatisation processes; however, unlike in the West, it affected the commons, which were not only fewer in number, but also underdeveloped, vulnerable, and more informal. Moreover, locating the common(ality) at the centre of communist ideology in CEE countries in fact put an end to commons as an institution for collective action beyond the classic private–public divide for decades to come. Although all available resources were officially handed over to the people and pronounced common property, in practice they remained under strict control of the socialist government. Not even allotment gardens or housing cooperatives—exemplary urban commons in the Western nomenclature—were left out as exceptions to the rule. To label a somewhat intermediary form of management applied in CEE after WWII, Insa Theesfeld (2019) ushers the notion of pseudo-commons. Described as the fourth possible method of counteracting the tragedy of the commons—alongside Hardin's (1968) alternatives of state regulation and privatisation and Ostrom's self-organisation of commoners—it involves appointing a resource manager in charge of the common resources on the part of the actual commoners. In state socialism, this would mean a civil servant or organisation representing the interests of citizens and not the state itself. Theesfeld discusses well-documented abuse of such arrangements, amounting to the appropriation of the commoners' rights and blurring of the responsibilities. She uses the example of *leskhozy*—Soviet agencies designed as community-based organisations for forest management across the USSR, which fell under the direct jurisdiction of the Communist Party. Therefore, the use of the prefix 'pseudo' not only stands for the artificiality of a *de-facto* top-down bureaucratic management that has little to do with genuine bottom-up commoning, but, perhaps more importantly, it points to the consequent erosion of institutions of social trust and cooperation. As Theesfeld argues (2019: 357), the 'vicious cycle of pseudo-collective action', in which the lack of trust inhibits cooperation and, *vice versa*, the lack of cooperation further increases mistrust, may have been the major factor hampering the emergence of common-property regimes after 1989.

While the uncommonness of commoning practices in cities after socialism remains a challenging research problem (Toto et al. forthcoming), the concept of urban commons in post-socialism has been embraced in a growing number of studies (Łapniewska 2017, Grabkowska 2018, Čukić and Timotijević 2020, Dellenbaugh-Losse et al. 2020, Matysek-Imielińska 2020, Toto et al. 2021). At present, commoning appears as just one of the many domains in which post-socialist cities seem to be following, or keeping up with, the global trends. However, even if today's urban communities of CEE seem to be heading in the same direction as their Western counterparts (which accounts for Bodnár's Unified Urban Logic), their road to modern commoning has been quite distinct and definitely bumpier (Socialist versus Capitalist Urban Logic). Furthermore, the journey itself had an altogether different starting point (Historical Continuity of Idiosyncratic Eastern Features).

The question remains what the term 'post-socialist' actually signifies, for just like there is no universal definition of neoliberalism, the post-socialist city comes in many shapes, sizes, and interpretations. Nonetheless, as demonstrated in Chapter 2, despite their variety, urban microcosms of Central and Eastern Europe (cf. Davies and Moorhouse 2002) share enough features and settings for the specific post-socialist urban common good to develop.

To escape the risk of reducing post-socialism to either a spatio-temporal location (a container) or hybridity of past and present (a condition), I adopt its de-territorialised understanding, as advocated by Tauri Tuvikene (2016: 133). By considering post-socialist cities not as a whole, similar entities, but rather with regard to their specific aspects—such as the urban common good—this approach admits the existence of crucial differences between them.

Despite their usually long and twisted histories, Central and Eastern European cities today are still mostly perceived through the prism of the 20th century, when they were state-controlled and governed by ideology. During this time, as will be elaborated in detail, the idea of the common good at the local level flourished at first, only to become politically sabotaged—the official mottos of egalitarianism and social justice turned out to be empty slogans. After communism had failed to live up to its name, its principles were compromised, and the rules for organising and sharing urban space were to be rewritten. But in the course of the transition from centrally planned to a market economy and from illiberal to liberal democracy, the conditions for the urban common good changed in an unforeseen way. While post-socialist cities approximated Western paradigms of the city as a commodity and as a commons, these were both modulated by the socialist legacy. As pointed out by Kimberly Zarecor (2018: 2), the socialist city 'does not exist only in the historic imagination' but 'still resonates in the present'.

By way of conclusion, before moving on to Chapter 2, I think of this book as seeking common ground between the post-socialist perspectives on urban common good. A common ground in the sense of an overarching idea, as in the central theme of the 13th Architectural Biennale in Venice. To paraphrase its curator David Chipperfield's (2012: 14) reflections on architecture, I find that urban common good

> requires collaboration, and most importantly it is susceptible to the quality of this collaboration. It is difficult to think of another peaceful activity that draws on so many diverse contributions and expectations. It involves commercial forces and social vision; it must deal with the wishes of institutions and corporations and the needs and desires of individuals. […] it requires a conspiracy of circumstances and participants.

My intention is thus to look inside the CEE pavilions of urban commonality, explore their multiplex content and make sense of the architecture of the common good, relating both to theory and practice.

Notes

1 Other concepts in this vein include the notion of plug-in citizen (Nawratek 2011) and cityzenship (Vrasti and Dayal 2016).

2 Together, they complement and support John Friedmann's (2000) four pillars of the socio-material framework of the good city: adequate housing, affordable health care, justly remunerated work, and sufficient provision of social services.

3 The reason for Asen's reluctance towards the word 'common' lies in his rejection of the entangled connotations of 'common content, a shared set of procedures, or a consensus-driven outcome' which, according to the author, stand in the way of public deliberation and recognition of difference in the process of **achieving** common good (336).

4 In her book on the power of hope as a motivation for collective action, Rebecca Solnit (2016) designates 9 November 1989 as the earliest of the five caesuras marking the coming of the new millennium. The other four are: 1 January 1994 (Zapatista uprising against NAFTA), 30 November 1999 (mass protests against WTO in Seattle), 11 September 2001 (al-Qaeda terrorist attacks against the United States), and 15 February 2003 (global demonstrations against the Iraq War).

References

Acemoglu D, Robinson JA (2012) *Why Nations Fail: The Origins of Power, Prosperity, and Poverty*. New York: Crown Business.

Amin A (2006) The Good City. *Urban Studies*, 43(5/6), 1009–1023.

Amin A, Howell P (2016) Thinking the commons. In: Amin A, Howell P (eds), *Releasing the Commons: Rethinking the Futures of the Commons*. London and New York: Routledge, 1–17.

Asen R (2017) Neoliberalism, the public sphere, and a public good. *Quarterly Journal of Speech*, 103(4), 329–349. https://doi.org/10.1080/00335630.2017.1360507

Asen R (2018) Introduction: Neoliberalism and the public sphere. *Communication and the Public*, 3(3), 171–175. https://doi.org/10.1177/2057047318794687

Attoh KA (2011) What kind of right is the right t the city? *Progress in Human Geography*, 35(5), 669–685.

Barbero I (2015) When rights need to be (re)claimed: Austerity measures, neoliberal housing policies and anti-eviction activism in Spain. *Critical Social Policy*, 35(2), 270–280. https://doi.org/10.1177/0261018314564036

Bauman Z (2003) *Liquid Love: On the Frailty of Human Bonds*. Cambridge: Polity.

Berge E, McKean M (2015) On the commons of developed industrialized countries. *International Journal of the Commons*, 9(2): 469–485.

Blandy S, Lister D (2005) Gated Communities: (Ne)Gating Community Development? *Housing Studies*, 20(1), 287–301.

Blokland T, Hentschel Ch, Holm A, Lebuhn H, Margalit T (2015) Urban citizenship and right to the city: The fragmentation of claims. *International Journal of Urban and Regional Research*, 39(4), 655–665, https://doi.org/10.1111/1468-2427.12259

Blomley N (2005) Flowers in the bathtub: boundary crossings at the public–private divide. *Geoforum*, 36, 281–296.

Bodnár J (2001) *Fin de Millenaire Budapest: Metamorphoses of Urban Life*. Minneapolis: University of Minnesota Press.

Bodnar J (2015) Reclaiming public space. *Urban Studies*, 52(12), 2090–2104.

Boggs C (1997) The great retreat: decline of the public sphere in late twentieth-century America. *Theory and Society*, 26, 741–80.

Bollier D (2014) *Think Like a Commoner. A Short Introduction to the Life of the Commons*. Gabriola Island: New Society Publishers.

Boniburini I (2013) La ville comme bien commun: Planification urbaine et droit à la ville. *Les Cahiers d'architecture La Cambre-Horta*, 9.

Borch Ch, Kornberger M (2015) *Urban Commons: Rethinking the City*. Abingdon and New York: Routledge.

Bravo G, De Moor T (2008) The commons in Europe: from past to future. *International Journal of the Commons*, 2(2), 155–161.
Brenner N, Theodore N (2002) Cities and the geographies of 'Actually Existing Neoliberalism'. In: Brenner N, Theodore N (eds), *Spaces of Neoliberalism: Urban Restructuring in North America and Western Europe*. Blackwell: Malden, 2–32.
Brenner N, Marcuse P, Mayer M (eds) (2011) *Cities for People, Not for Profit: Critical Urban Theory and the Right to the City*. London: Routledge.
Budds J, Teixeira P (2005) Ensuring the Right to the City: Pro-Poor Housing, Urban Development and Tenure Legalization in São Paulo, Brazil. *Environment and Urbanization*, 17(1), 89–113.
Cabannes Y, Delgado C (ed.) (2015) *Participatory Budgeting, Dossier No 1*. Another city is possible! Alternatives to the city as a commodity series, Lisbon.
Calhoun C (ed.) (1993) *Habermas and the Public Sphere*. Cambridge: MIT Press.
Carlos AFA, Volochko D, Pinto Alvarez I (2017) *The city as a commodity*. São Paulo: Faculty of Philosophy, Letters and Human Sciences of the University of São Paulo. https://doi.org/10.11.606/9788575063026
Carmona M (2015) Re-theorising contemporary public space: a new narrative and a new normative, *Journal of Urbanism: International Research on Placemaking and Urban Sustainability*, 8(4), 373–405. https://doi.org/10.1080/17549175.2014.909518
Castells M (1977) *The Urban Question. A Marxist Approach*. London, Edward Arnold.
Chelcea L, Druţă O (2016) Zombie Socialism and the Rise of Neoliberalism in Post-Socialist Central and Eastern Europe. *Eurasian Geography and Economics*, 57(4–5), 521–544.
Chipperfield D (2012) Introduction. In: Chipperfield D, Long K, Bose S (eds), *Common Ground: A Critical Reader*. Venice: Marsilio Editori, 14.
Colby CC (1933) Centrifugal and centripetal forces in urban geography. *Annals of the Association of American Geographers*, 23(1), 1–20. https://doi.org/10.1080/00045603309357110
Colomb C, Novy J (eds) (2017) *Protest and Resistance in the Tourist City*. Abingdon: Routledge.
Čukić I, Timotijević J (eds) (2020) *Spaces of Commoning. Urban Commons in the Ex-Yu Region*. Belgrade: Ministry of Space/Institute for Urban Politics.
Czepczyński M (2008) Cultural *Landscapes of Post-Socialist Cities: Representation of Powers and Needs*. Aldershot: Ashgate.
Dale G, Fabry A (2018) Neoliberalism in Eastern Europe and the Former Soviet Union. In: Cahill D, Cooper M, Konings M, Primrose D (eds), *The SAGE Handbook of Neoliberalism*. London: SAGE, 234–247.
Dasgupta P, Ramanathan V (2014) Pursuit of the common good. *Science*, 345(6203), 1457–1458. https://doi.org/10.1126/science.1259406
Davies N, Moorhouse R (2002) *Microcosm: Portrait of a Central European City*. London: Jonathan Cape.
De Angelis M (2007) *The Beginning of History*. London: Pluto Press.
Dear M (1992) Understanding and Overcoming the NIMBY Syndrome. *Journal of the American Planning Association*, 58(3), 288–300.
Dear M (2011) The Urban Question after Modernity. In: Schmid H, Sahr W-D, Urry J (eds.), *Cities and Fascination Beyond the Surplus of Meaning*. Burlington: Ashgate, 17–31.
Dellenbaugh M, Kip M, Bieniok M, Müller AK, Schwegmann M (red.) (2015) *Urban Commons: Moving Beyond State and Market*. Basel: Birkhäuser.

Dellenbaugh-Losse, M., Zimmermann, N. and De Vries, N., 2020. The Urban Commons Cookbook: Strategies and Insights for Creating and Maintaining Urban Commons.

Domaradzka A (2018) Urban Social Movements and the Right to the City: An Introduction to the Special Issue on Urban Mobilization. *Voluntas*, 29, 607–620.

Eijk Van G (2010) Exclusionary Policies are Not Just about the 'Neoliberal City': A Critique of Theories of Urban Revanchism and the Case of Rotterdam. *International Journal of Urban and Regional Research*, 34(4), 820–834.

Enright TE (2013) Mass Transportation in the Neoliberal City: The Mobilizing Myths of the Grand Paris Express. *Environment and Planning A: Economy and Space*, 45(4), 797–813. https://doi.org/10.1068/a459

Enright T, Rossi U (2018) Ambivalence of the Urban Commons. In: Ward K, Jonas AEG, Miller B, David Wilson D (eds.), *The Routledge Handbook on Spaces of Urban Politics*. Abingdon: Routledge, 35–46.

Enyedi G (1996) Urbanization under socialism. In: G Andrusz, M Harloe, I Szelenyi (eds), *Cities after Socialism: Urban and* Regional *Change and Conflict in Post-socialist Societies*. Oxford: Blackwell, 100–118.

Fahmi WS (2009) Bloggers' street movement and the right to the city. (Re)claiming Cairo's real and virtual "spaces of freedom". *Environment and Urbanization*, 21(1), 89–107.

Fainstein S (2010) *The Just City*. Ithaca: Cornell University Press.

Fairclough N (2001) *Language and Power*. London: Routledge.

Fennell LA (2011) Ostrom's Law: Property Rights in the Commons. *International Journal of the Commons*, 5(1), 9–27.

Fenton N (2018) Fake democracy: the limits of public sphere theory. *Javnost: The Public*, 25(1–2), 28–34, https://doi.org/10.1080/13183222.2018.141882

Fenton N, Downey J (2003) Counter Public Spheres and Global Modernity. *Javnost - The Public*, 10(1), 15–32. https://doi.org/10.1080/13183222.2003.11008819

Ferenčuhová S (2012) Urban theory beyond the 'East/West divide'? Cities and urban research in postsocialist Europe. In: Edensor T, Jayne M (eds), *Urban Theory beyond the West: A World of Cities*. London and New York: Routledge, 65–74.

Foster SR, Iaione Ch (2016) The City as a Commons. *Yale Law & Policy Review*, 34(2), 282–349.

Fraser N (1996) Rethinking the public sphere: A contribution to the critique of actually existing democracy. In: Calhoun C (ed.), *Habermas and the Public Sphere*. Cambridge: Massachusetts Institute of Technology, 109–142.

Friedmann J (2000) The good city: in defense of utopian thinking. *International Journal of Urban and Regional Research*, 24(2), 460–472.

Friendly A (2013) The right to the city: theory and practice in Brazil. *Planning Theory and Practice*, 14(2), 158–179.

Gibson-Graham JK, Cameron J, Healy S (2016) Commoning as a postcapitalist politics. In: A Amin, P Howell (eds.), *Releasing the Commons: Rethinking the Futures of the Commons*. London and New York: Routledge, 192–212.

Graeber D (2013) *The Democracy Project: A History, a Crisis, a Movement*. New York: Spiegel & Grau.

Grabkowska M (2018) Urban Space as a Commons in Print Media Discourse in Poland after 1989. *Cities*, 72, 122–129.

Guarneros-Meza V, Geddes M (2010) Local governance and participation under neoliberalism: Comparative perspectives. *International Journal of Urban and Regional Research*, 34(1), 115–129.

Haas T, Olsson K (2014) Transmutation and reinvention of public spaces through ideals of urban planning and design. *Space and Culture*, 17(1), 59–68.

Habermas J (1989) *The Structural Transformation of the Public Sphere: An Inquiry into a Category of Bourgeois Society*. Cambridge: MIT Press.

Hardin G (1968) The tragedy of the commons: the population problem has no technical solution; it requires a fundamental extension in morality. *Science*, 162, 1243–1248.

Hartshorne R (1950) The functional approach in political geography. *Annals of the Association of American Geographers*, 40(2), 95–130.

Hardt M, Negri A (2009) *Commonwealth*. Cambridge: Harvard University Press.

Harvey D (1973) *Social Justice and the City*. London: Arnold.

Harvey D (1989) From managerialism to entrepreneurialism: The transformation of urban governance in late capitalism. *Geografiska Annaler, Series B, Human Geography*, 71(1), 3–17. https://doi.org/10.2307/490503

Harvey D (2003) The right to the city. *International Journal of Urban and Regional Research*, 27(4), 939–941.

Harvey, David (2007a). *A brief history of neoliberalism*. Oxford University Press. ISBN 978-0199283279.

Harvey D (2007b) Neoliberalism as creative destruction. *Annals of the American Academy of Political and Social Science*, 610, 28–29.

Harvey D (2008) The right to the city. *New Left Review*, 53, 23–40.

Harvey D (2012) *Rebel Cities: From the Right to the City to the Urban Revolution*. London: Verso.

Healey P (2002) On creating the 'city' as a collective resource. *Urban Studies*, 39(10), 1777–1792.

Hess Ch, Ostrom E (2007) Introduction: An overview of the knowledge commons. In: Hess Ch, Ostrom E (eds.), *Understanding Knowledge as Commons: From Theory to Practice*. Cambridge: MIT Press, 3–26.

Hörschelmann K, Stenning A (2008) Ethnographies of postsocialist change. *Progress in Human Geography*, 32(3), 339–361.

Howe E (1992) Professional roles and the public interest in planning. *Journal of Planning Literature*, 6(3), 230–248.

Hudson B, Rosenbloom J, Cole D (eds.) (2019) *Routledge Handbook of the Study of the Commons*. London and New York, Routledge.

Kideckel DA (2002) The unmaking of an east-central European working class. In: Hann C (ed), *Post-socialism: Ideals, ideologies and practices in Eurasia*. London: Routledge, 114–132.

Krakovský R (2018) *State and Society in Communist Czechoslovakia: Transforming the Everyday from World War II to the Fall of the Berlin Wall*. London: I.B. Tauris.

Krastev I, Holmes S (2019) *The Light that Failed: A Reckoning*. London: Allen Lane.

Koczanowicz L (2021) Common goods or common interests: two visions of democratic society. *Pragmatism Today*, 12(2), 198–204.

Krugman P (1994). Urban concentration: The role of increasing returns and transport costs. *The World Bank Economic Review*, 8(suppl 1), 241–263. https://doi.org/10.1093/wber/8.suppl_1.241

Laborda Pemán M, De Moor T (2013) A Tale of Two Commons. Some Preliminary Hypotheses on the Long-term Development of the Commons in Western and Eastern Europe, 11th–19th Centuries. *International Journal of the Commons*, 7(1), 7–33.

van Laerhoven FSJ, Barnes CA (2014) Communities and commons: The role of community development support in sustaining the commons. *Community Development Journal*, 49(S1), i118–i132.
van Laerhoven F, Ostrom E (2007) Traditions and trends in the study of the Commons. *International Journal of the Commons*, 1(1), 3–28.
Łapniewska Z (2017) (Re)claiming Space by Urban Commons. *Review of Radical Political Economics*, 49(1), 54–66.
Lees L, Shin HB, López-Morales E (eds) (2015) *Global Gentrifications: Uneven Development and Displacement*. Bristol: Policy Press.
Lefebvre H (1968) *Le Droit à la Vill*e. Anthropos: Paris.
Lefebvre H (1974) *La production de l'espace*. Anthropos: Paris.
Lefebvre H (1996) *Writings on Cities*. Wiley-Blackwell.
Lenna V, Trimarchi M (2019) For a culture of urban commons: practices and policies. In: Benincasa C, Neri G, Trimarchi M (eds), *Art and Economics in the City*. Bielefeld: transcript Verlag, 205–244.
Linebaugh P (2008) *The Magna Carta Manifesto: Liberties and Commons for All*. Berkeley: University of California Press.
Marcuse P (2009) From critical urban theory to the right to the city. *City*, 13(2–3), 185–197.
Marhold H (2015) A European Common Good? *L'Europe en Formation*, 2, 9–24.
Massey D (2005) *For Space*. Los Angeles: Sage.
Matysek-Imielińska M (2020) *Warsaw Housing* Cooperative*: City in Action*. Cham: Springer.
Mayer M (1999) Urban Movements and Urban Theory in the Late-20th-Century City. In: Beauregard RA, Body-Gendrot S (eds) *The Urban Moment: Cosmopolitan Essays on the Late-20th-Century City*. Thousand Oaks: Sage Publications, 209–238.
Mayer M (2014) Toward Glocal Movements? New Spatial Politics for a Just City. In: Eick V, Briken K (eds.), *Urban (In)security: Policing the Neoliberal Crisis*. Ottawa: Red Quill Books, 27–59.
Mayer M (2018) Neoliberalism and the Urban. In: Cahill D, Cooper M, Konings M, Primrose D (eds), *The SAGE Handbook of Neoliberalism*, London: SAGE, 483–495.
Merrifield A (2013) *The Politics of the Encounter: Urban Theory and Protest under Planetary Urbanization*. Athens: University of Georgia Press.
Merrifield A (2014) *The New Urban Question*. London: Pluto Press.
Miles M (2009) Public Spheres. In: Harutyunyan A, Hörschelmann K, Miles M (eds) *Public Spheres After Socialism*. Bristol: Intellect Books, 133–150.
Montgomery C (2013) *Happy City: Transforming Our Lives Through Urban Design*. ....
Moss T (2014) Spatiality of the commons. *International Journal of the Commons*, 8(2): 457–471.
Mouffe C (2000) *The Democratic Paradox*. London: Verso.
Mulíček O (2009) Prostorové vzorce postindustriálního Brna. In: Ferenčuhová S, Galčanová L, Hledíková M, Vacková B (eds), *Město: Proměnlivá nelsamozřejmost*. Červený Kostelec/Brno: Pavel Mervart/Masarykova Univerzita, 53–175.
Murphy E, Fox-Rogers L (2015) Perceptions of the common good in planning. *Cities*, 42, 231–241.
Nawratek K (2011) *City as a Political Idea*. Plymouth: University of Plymouth.

Negt O, Kluge A (1993) *Public Sphere and Experience: Toward an Analysis of the Bourgeois and Proletarian Public Sphere*. Minneapolis: University of Minnesota Press.

Németh J (2009) Defining a public: The management of privately owned public space. *Urban Studies*, 46(11): 2463–2490.

New Leipzig Charter: The transformative power of cities for the common good (2020) https://ec.europa.eu/regional_policy/en/newsroom/news/2020/12/12-08-2020-new-leipzig-charter-the-transformative-power-of-cities-for-the-common-good

Ostrom E (1990) *Governing the Commons: The Evolution of Institutions for Collective Action*. Cambridge: Cambridge University Press.

Pobłocki K (2017) *Kapitalizm: historia krótkiego trwania*. Warszawa: Fundacja Bęc Zmiana.

Pickles J, Smith A (eds) (1998) *Theorising Transition: The Political Economy of Post-communist Transformations*. London: Routledge.

Peck J, Theodore N, Brenner N (2012) Neoliberalism resurgent? Market rule after the great recession. *South Atlantic Quarterly*, 111(2), 265–288.

Peck J (2013) Explaining (with) Neoliberalism. *Territory, Politics, Governance*, 1(2), 132–157. https://doi.org/10.1080/21622671.2013.785365

Petsinis V (2010) Twenty years after 1989: moving on from transitology. *Contemporary Politics*, 16(3), 301–319.

Pickvance C (2003) From urban social movements to urban movements: a review and introduction to a symposium on urban movements. *International Journal of Urban and Regional Research*, 27(1), 102–109.

Plyushteva A (2009) The right to the city and struggles over urban citizenship: Exploring the links. *Amsterdam Social Science*, 1(3): 81–97.

Portney KE (2013) *Taking Sustainable Cities Seriously: Economic Development, the Environment, and Quality of Life in American Cities*. Cambridge: MIT Press.

Purcell M (2002) Excavating Lefebvre: The right to the city and its urban politics of the inhabitant. *GeoJournal*, 58, 99–108.

Rawls J (1971) *A Theory of Justice*. Cambridge: Belknap Press.

Rogge N, Theesfeld I (2018) Categorizing urban commons: Community gardens in the Rhine-Ruhr agglomeration, Germany. *International Journal of the Commons*, 12(2), 251–274.

Ryder J (2013) *The Things in Heaven and Earth: An Essay in Pragmatic Naturalism*. New York: Fordham University Press.

Sæter O (2011) The body and the eye: Perspectives, technologies, and practices of urbanism. *Space and Culture*, 14(2) 183–196.

Samara TR, He S, Chen G (eds) (2013) *Locating right to the city in the global south*. London: Routledge.

Sandberg A, Theesfeld I, Schlüter A, Penov I, Dirimanova V (2013) Commons in a changing Europe. *International Journal of the Commons*, 7(1), 1–6.

Sandel M (1982) *Liberalism and the Limits of Justice*. Cambridge: Cambridge University Press.

Schively C (2007) Understanding the NIMBY and LULU Phenomena: Reassessing Our Knowledge Base and Informing Future Research. *Journal of Planning Literature*, 21(3), 255–266.

Sevilla-Buitrago A (2015) Crisis and the city: neoliberalism, austerity planning and the production of space. In: Eckardt F, Ruiz J (eds.), *City of Crisis: The Multiple Contestation of Southern European Cities*. Bielefeld: 31–49.

Shields R (2013) Lefebvre and the right to the open city? *Space and Culture*, 16(3) 345–348. https://doi.org/10.1177/1206331213491885
Smith TW (1999) Aristotle on the conditions for and limits of the common good. *American Political Science Review*, 93(3), 625–636.
Soja EW (1989) *Postmodern Geographies: The Reassertion of Space in Critical Social Theory*. London: Verso.
Soja EW (2010) *Seeking spatial justice*. Minneapolis: University of Minnesota Press.
Staeheli LA, Mitchell D (2007) Locating the public in research and practice. *Progress in Human Geography*, 31(6): 792–811.
Stavrides S (2016) *Common Space: The City as Commons*. London: Zed Books.
Storper M (2016) The neo-liberal city as idea and reality. *Territory, Politics, Governance*, 4(2), 241–263.
Strathern M (2016) Inroads into altruism. In: A Amin, P Howell (eds.), *Releasing the Commons: Rethinking the Futures of the Commons*. London and New York: Routledge, 161–176.
Szelényi I (1996) Cities under socialism – And after. In: Andrusz G, Harloe M, Szelenyi I (eds), *Cities after Socialism: Urban and Regional Change and Conflict in Post-socialist Societies*. Oxford: Blackwell, 286–317.
Tait M (2011) Trust and the public interest in the micropolitics of planning practice. *Journal of Planning Education and Research*, 31(2), 157–171.
Taylor C (1989) *Sources of the Self: The Making of Modern Identity*. Cambridge: Harvard University Press.
Theesfeld I (2019) The role of pseudocommons in post-socialist countries. In: Hudson B, Rosenbloom J, Cole D (eds), *Routledge Handbook of the Study of the Commons*. London and New York: Routledge, 345–359.
Tonkiss F, Passey A, Fenton N, Hems L (eds) (2000) *Trust and Civil Society*. London: Palgrave Macmillan.
Toto R, Ćaćić M, Ivanova Z, Nientied P, Stachowiak-Bongwa K (2021). Urban Commons in Emerging Economies. In: Fransen J, van Dijk MP, Edelenbos J (eds) *Urban Planning, Management and Governance in Emerging Economies. New Paradigms and Practices*. Cheltenham: Edward Elgar, 117–141.
Tummers L (2016) The re-emergence of self-managed co-housing in Europe: A critical review of co-housing research. *Urban Studies*, 53(10), 2023–2040.
Tuvikene T (2016) Strategies for Comparative Urbanism: Post-socialism as a De-territorialized Concept. *International Journal of Urban and Regional Research*, 40(1), 132–146.
Unnikrishnan H, Manjunatha B, Nagendra H (2016) Contested urban commons: mapping the transition of a lake to a sports stadium in Bangalore. *International Journal of the Commons*, 10(1), 265–293.
Vrasti W, Dayal S (2016) Cityzenship: rightful presence and the urban commons, *Citizenship Studies*, 20(8), 9941011. https://doi.org/10.1080/13621025.2016.1229196
Wiktionary (2022) Retrieved from https://en.wiktionary.org/wiki/share (last accessed: 18 February 2022).
Wollmann H (2006) The fall and rise of the local community: A comparative and historical perspective. *Urban Studies*, 43(8), 1419–1438.
Wong TC, Liu R (2017) Developmental urbanism, city image branding and the "Right to the City" in transitional China. *Urban Policy and Research*, 35(2), 210–223.
Wu F (2017) State entrepreneurialism in urban China. In: Pinson G, Journel CM (eds.), *Debating the Neoliberal City*. Abingdon: Routledge, 153–173.
Zarecor KE (2018) What Was So Socialist about the Socialist City? Second World Urbanity in Europe. *Journal of Urban History*, 44(1), 95–117.

2 Transforming conceptions of urban common good in Central and Eastern Europe

Urban common good during and after socialism

While the features of urban commonality presented in Chapter 1 apply to cities in general, reconfigurations of the model of urban common good are more context-dependent and, therefore, less universal. Although the transformations observed in Central and Eastern Europe resemble Western European patterns in many ways, they also vary substantially in certain aspects. First of all, it must be emphasised that east of the Elbe the origins of the modern public sphere did not fully align with the Habermasian concept. Due to the durability of feudalism in CEE and the region's political instability, publicness developed only slowly and gradually (Nowak and Pluciński 2011). Moreover, depending on the country, it was either detached from the urban milieus or lingered on the periphery of urban culture. The same conditions that hampered the development of the commons in Eastern Europe explain the relative backwardness of CEE with regard to urbanity (Kubicki 2019). Even in 19th-century Silesia, one of the most economically advanced nooks of the region, the extraordinary industry-led expansion of cities was not accompanied by flourishing urban culture or bourgeois public sphere. In comparison with the west of Europe, the bourgeoisie's efforts towards self-determination were marginal across the region. In Poland, where this social group was too weak to 'become a leading force and organise national consciousness' (Gömöri 1973: 154), the central position in the emerging public sphere was taken over by intelligentsia of gentry descent, whose system of cultural values was not exactly city-oriented, if not explicitly anti-urban (Kubicki 2011). The absence of the bourgeois ethos as a vital social drive prevented the urban common good from being considered a matter of public interest or concern. Therefore, the collective actors of the early capitalist CEE public sphere formed a pseudo-public, resembling the transformed bourgeois model but deprived of its urban character. This further contributed to the barrenness of public spirit in developing cities, which remained sites of consumption more than of debate.

Another departure from the conditions described by Habermas concerns the contested sovereignty of CEE states on the threshold of modernity. Unlike their robust and globally competitive Western European counterparts,

DOI: 10.4324/9781003089766-4

countries in the East were relatively vulnerable and unstable, both in political and economic terms. The incorporation of the Polish-Lithuanian Commonwealth and the Kingdom of Bohemia into the neighbouring empires left them practically devoid of independence. The former case involved over a 100 years of annexation, following three partitions at the end of the 19th century. During that time, the authoritarian rule of the three invaders—the Kingdom of Prussia, the Russian Empire, and Habsburg Austria—instilled the perception of the 'public' as oppressive and dependent on state authority and power, thus additionally hindering the emergence of a public sphere in the Habermasian sense:

> Resentment towards the Leviathan [...] was not a discursive contestation leading to a 'civic recovery' of the state, but a form of open hostility towards the foreign. Since the Polish state had been disowned, there were no positive projects either, no perspectives for setting up its own public sphere.
>
> (Nowak and Pluciński 2011: 43)

A brief period of regained state autonomy between the two world wars was not sufficient for the public spheres to crystallise in many countries of CEE, and this facilitated the installation of totalitarian communist regimes after 1945. With the new order came a new kind of publicness, 'obviously shaped by political considerations and designed to serve primarily political goals' (Jakubowicz 1993: 169). Researchers have proposed numerous, and at times conflicting, takes on the common characteristics of the state socialist public sphere and its variations across countries. Some definitions underline its undemocratic disposition and orientation toward production (Cymbrowski 2017), while others mainly pay attention to the radical merging of the private and the public (Krakovský 2018). However, as maintained by Ingrid Oswald and Viktor Voronkov (2004: 104), 'a broad consensus exists that [it] has definitely *not* been akin to a forum of private citizens assembling as an audience outside (or against) the state'. Thus, quite ironically, the state-controlled public sphere in people's republics did not leave much space for deliberation. With a certain dose of sarcasm, Roman Krakovský (2018: 105) remarks that '[i]nstead of being a space for the exchange of opinions on matters relating to the common good, the socialist public sphere was largely used to mobilise and inform the population after decisions had already been made'. Malcolm Miles (2009: 133) even goes as far as to call it 'the state itself, with its apparatus of committees and other organs, including the security service'.

Until the mid-1950s, in CEE there had been no alternative to the state socialist public sphere as a conveyor of the dominating official discourse. However, during the Thaw period following Stalin's death, it was possible for counterpublics to appear in most countries of the region. While some accept Negt and Kluge's argument (1993) that these counterpublics were ephemeral, surfacing only momentarily at times of upheaval and protest, others see them in terms of continuous processes of evolution or coexistence (Figure 2.1).

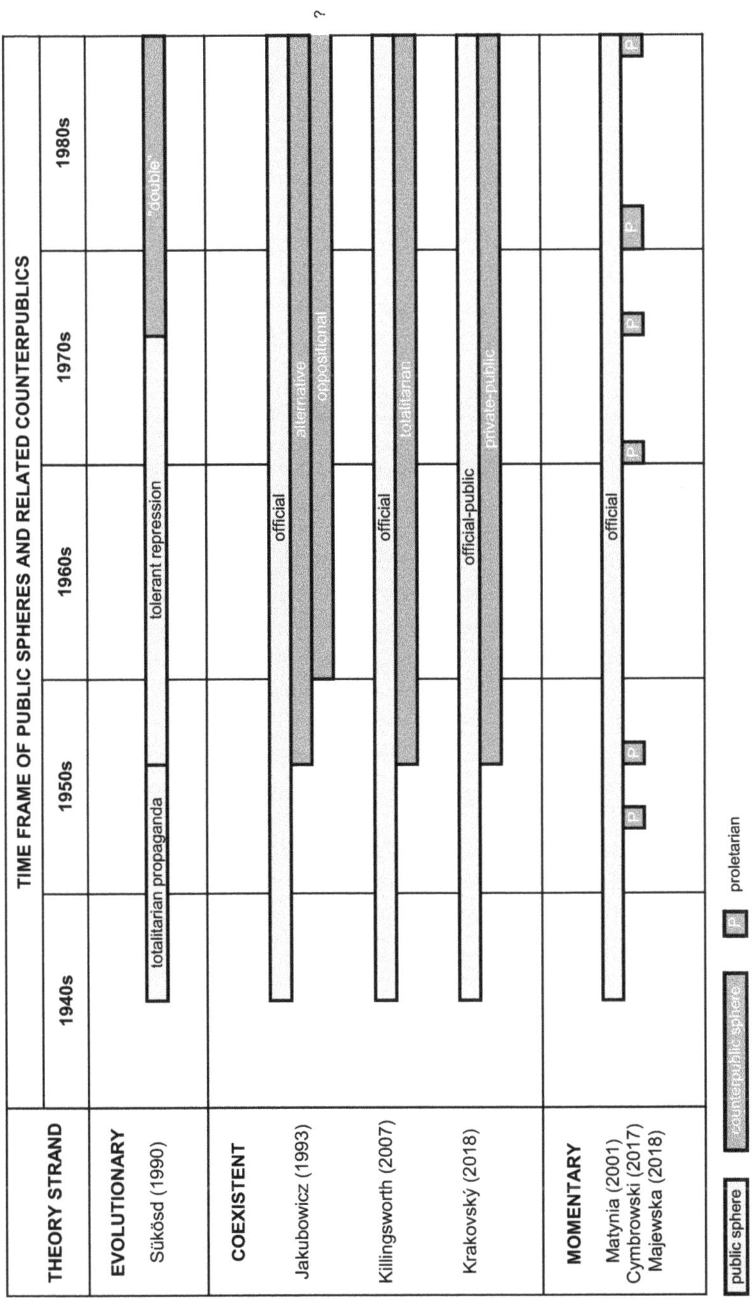

Figure 2.1 Chronological classification of public and counterpublic spheres in Central and Eastern Europe under state socialism.

Representing the evolutionary outlook, Miklós Sükösd (1990) distinguishes four consecutive phases of transformation of the public sphere: the totalitarian propaganda state, 'tolerant repression', the 'double' public sphere, and the post-communist public sphere. The first marks the post-war decade and therefore the most repressive period. During that time, the public sphere equalled a media system 'owned and controlled by the state or directly by the party', in which 'any form of public discourse, from political debates to entertainment and theoretical debates, became defined and dominated by communist authorities' (Sükösd 1990: 41). Its loosening into 'tolerant repression' occurred after the anti-communist protests that swept through the region in the mid-1950s and 1960s. In Czechoslovakia, the events of the Prague Spring put an end to the monopoly of propaganda, allowing the publication of non-censored political content, whereas in other countries the censorship became less pervasive. This shift was accompanied by a cultural change brought about by technological modernisation; it consisted of the liberalisation of the elite culture and partial de-ideologisation of mass culture.

Around the mid-1970s, the restrictions were further eased in Poland, Czechoslovakia, and Hungary, which, along with the expansion of the independent communication means, enabled the constitution of a non-official counterpublic. The resulting 'double' public sphere denotes a situation in which the alternative, 'second' public sphere, bottom-up in origin and civic in character, mirrors the 'first', state-imposed, while operating outside the state's control. As stressed by Sükösd, the key feature of this phase is proliferation of clandestine publishing, which leads to the 'crystallisation of oppositional activities' (Sükösd 1990: 54). In this manner, the 'genuine' counterpublic emulates the 'artificial' official public sphere by means of an apparent critical counter-response. Therefore, in comparison, it appears to come closer to the Habermasian ideal.

This view is taken to the extreme by Matt Killingsworth (2007), who puts the alternative public sphere before the official one and labels it totalitarian. His work represents the conceptual strand which I refer to as **coexistent**. Instead of duplication of the official public sphere, it assumes the formation of one or more parallel counterpublics, running independently. For example, Karol Jakubowicz (1993) identifies three competing public spheres in communist Poland: the official, the alternative, and the oppositional. The author attributes the emergence of the alternative public sphere, rooted in the milieu of the Catholic Church, to the already mentioned weakening of totalitarian control in the mid-1950s. This, in turn, paved the way for the oppositional public sphere, connected to the civic ferment which eventuated in the Solidarity movement. Although until the 1980s the magnitude of the two unofficial public spheres was limited, in the end, they turned out as 'powerful instruments of opinion- and will-formation' (Jakubowicz 1993: 159).

Finally, the third approach—**momentary** counterpublics—accounts for the state-socialist version of the proletarian public sphere under capitalism proposed by Negt and Kluge (1993). As it spotlights the moments of workers' mobilisation, it is rather discontinuous and ephemeral, or literally momentary (Cymbrowski 2017), unlike the alternative public spheres discussed above. In the Polish context, Elżbieta Matynia (2001) limits its existence to the sixteen months between the signing of the Gdańsk Agreement on 31 August 1980 and the introduction of the martial law on 13 December 1981. According to Matynia (2001: 930), the then-born 'morally sensitive public sphere' was enacted through 'a debate on collective identity that strongly emphasized the commonality of values within the res publica'. This argument is backed by the official programme of the Independent Self-governing Trade Union 'Solidarność', elaborated by the 1st National Congress of Delegates (Program NSZZ 'Solidarność' 1981) and demanding 'democracy, truth, rule of law, human dignity, freedom of thought, the repair of the Republic, not just bread, butter and sausage'. While the protests in the Eastern Bloc between the 1950s and the 1990s were typically driven by employees in the state-owned industrial sector, in most instances they managed to rally broad support among other social groups. Matynia thus points out the inclusiveness of proletarian counterpublics as a stark difference from the Habermasian model of the public sphere. She accentuates the fact that 'it was actually the *people* who took over much of the discourse ordinarily carried on by the educated classes' (2001: 934). This contrast is further emphasised by Ewa Majewska:

> Nothing in this description follows Habermas's classical script. The workers were not bourgeoisie, their workplace was not a salon, their clandestine publications cannot be compared with the official oppositional press, and some of them were women. Nothing seems to fit. Habermas says openly in the preface to the *Structural Transformation of the Public Sphere* that plebeian public spheres were historically insignificant, and he would therefore ignore them.
>
> (2018: 240)

One important analogy between the bourgeois and proletarian public spheres is that both thrived in the urban space. Just as sidewalk cafés were the sites of public debate for the bourgeoisie, the working class in the broad sense of the term—including students and intellectual authorities—enacted civic engagement in the streets. Cymbrowski (2017) hence notes that the common denominator uniting the actors of proletarian counterpublics in the Eastern Bloc was the city. Despite contextual differences, from the East German uprising of 1953 to the 'carnival of Solidarity' in 1980–1981, the proletarian publicness—produced 'either on the shop floors of factories or in public spaces of cities where those factories were located' (2017: 16)—was immersed in urban commonality. What makes it distinct from the capitalist proletarian public sphere is that it opposed the political and economic system of authoritarian state as a whole and not just the capitalist mode of production and property relations.

The conclusions that follow from this brief overview are twofold. Firstly, counterpublics were essential elements of public life under European state socialism in the second half of the 20th century. Whether from the evolutionary, coexistent, or momentary perspective, the alternative public sphere appears to have carried more weight under socialism than in capitalist countries of that time. Some authors (Matynia 2001, Killingsworth 2007) even contend that since the notion of civil society in the Western sense had not been applicable to the Eastern Bloc until 1989, studies of that period should employ the analytical framework of alternative public sphere instead.[1] Killingsworth considers the latter to be a '"steppingstone" in the evolution toward a possible fully functioning civil society' (2007: 76). Secondly, regardless of the theoretical strand, the CEE counterpublics had an especially firm urban grounding, which made them efficient Petri dishes for culturing bottom-up urban commonality in opposition to the sterile official publicness. Therefore, it may be argued that the alternatives to the state socialist public sphere had the potential not only to engender the forthcoming civil society, but also to lay the foundations for prospective practices and institutions of the urban common good.

At the turn of the 1990s, publicness in CEE underwent rapid reorientation. The process was facilitated by liberalisation and pluralisation of the media, even if their independence was impaired due to reliance on private capital and affiliation with new and former elites (Sula 2008, Bajomi-Lázár 2017). Whereas some forecasts envisaged a prompt adoption of the Western standards (Sükösd 1990), others were more cautious. For instance, Jakubowicz (1993) considered three scenarios of change. The first involved the expiration of the alternative public sphere as a result of a consensus on the shape of the transformed social order. According to the second one, a new official public sphere would crystallise along with a new counterpublic. In the last scenario, the dominant official public sphere would be accompanied by not one but several counterpublics, representing various dissenting groups (Jakubowicz 1993: 169). The inconclusiveness of these estimations reflects the socio-political and economic destabilisation of that time. It was apparent that the aftershocks of the seismic shift of 1989 would be felt for years to follow, generating greater uncertainty about the future.

More than a decade into the transition in Russia, Oswald and Voronkov (2004) assessed that no autonomous public sphere had surfaced there at all. They found that the citizens' relationship with the state congealed at the point where the former would confront the latter 'with all-encompassing expectations and demands' and, at the same time, 'distrust the state and refuse to engage in public affairs' (2004: 114).[2] Correspondent figures of homo sovieticus (Zinoviev 1986, Heller 1988, Tischner 1992) and Soviet mentality (Mikheyev 1987, Utekhin 2018) have been prominent in academic literature. In the western part of the former Eastern Bloc, the similarly low level of social trust and insensitivity to the common good translated into a set of such characteristics as a culture of mistrust (Sztompka 1996, Giordano and Kostova 2002), weakness of social dialogue (Mailand and Due 2004) and molecular development of society (Czapiński and Panek 2015). Yet, citizens of the former Soviet satellite countries had mobilised rather quickly to break away from the past and proceed to reclaim the official public sphere. In Poland, a brusque farewell to the compromised regime was bid with the use of the 'thick line' policy from the 1989 designation speech of Tadeusz Mazowiecki, the first non-communist Prime Minister in the Eastern Bloc after World War II (WWII) (Ornatowski 2008). Romanian intellectuals were even more radical, dismissing not only communism but the left in general, together with its system of values:

> From a neoliberal perspective, the left is rejected on account of its collectivism, statism, anti-individualism; from a conservative perspective, the left is rejected on account of its egalitarianism, progressivism and radicalism, as the embodiment of a pathological obsession with revolutionary change.
>
> (Preoteasa 2002: 274)

With this background, the newly post-socialist cities quickly followed the tracks of their old-capitalist counterparts in an accelerated transition to neoliberalism. As reasoned by Alison Stenning et al. (2010: 34), not only did the transformation from state-controlled to market-oriented economy involve mass privatisation and withdrawal of the state from public intervention, but also promoted 'individualism and individual responsibility for health, wealth and welfare'. A deep societal shift ensued, aggravated by the overall turbulence in the socio-economic sphere, with soaring inflation and unemployment rates.

The extraordinary pace of the process was not the sole difference. Drawing an analogy between Karl Polanyi's Great Transformation (1944) and the Central and Eastern European transition from socialism to capitalism, Judit Bodnár (2001) notes that in both cases one of the major factors of change consisted in the enclosure of the common space. Considering the predominantly rural character of the precapitalist 'common' and the more urban disposition of its socialist counterpart— resulting also from their temporal distance of more than two centuries—the privatisation and fragmentation processes in CEE had a relatively more profound impact on cities. With mixed feelings, Bodnár observes how the 'liberating aspects of the widening possibilities for the particularistic use of urban space, the turning inward of urban lives, and the concomitant losses of the declining significance of public goods and public space leave ambiguous markings on the recent metamorphosis of urban life (2001: 10).

This brings the discussion of the socialist and post-socialist publicness to the question of urban common good. Investigating the public sphere under communism from a local perspective, Krakovský (2018: 80) warns against focusing on the physical dimension alone and neglecting two other vital components: the nature of the citizen community and the functioning of its politics. This requirement is met by the model of urban common good introduced in Chapter 1. In keeping with Asen's (2017) triad, the spatial dimension of the public sphere parallels the urban paradigm, the nature of the citizen community is represented by collective human actors, and the political conditions are reflected in the dominant rationales. As in the Western European case, the reconfigurations of the model during and after state socialism stem from external and internal conditions, which affect the relationships between the three components and translate into the interplay of centripetal and centrifugal forces (Table 2.1).

Having introduced the early capitalist (or pre-socialist) public sphere and its urban dimension, I will now turn to an analysis of the model's dynamic under socialism, that is, from the mid-1940s to the 1990s, and after, that is, since the 1990s. The latter period is subdivided into two parts. It begins with what I refer to as the transition stage. Although the term is imprecise, I will use it to distinguish the first two decades of the post-1989 CEE transformation. A time of political, economic, and socio-cultural upheaval, it also saw the development of the post-socialist public sphere. The transition symbolically ends

Table 2.1 Transformations of the model of urban common good in Central and Eastern Europe

Type of public sphere	*Approximate timeline*	*Elements of urban common good*			*Dynamics of urban common good*
		collective human actors	*dominant rationales*	*urban paradigm*	
Early capitalist public sphere	19th century to mid-1940s	pseudo-public	privatism, consumerism	city as a space for consumption	centrifugal
State socialist-public sphere	mid-1940s to 1990s	'citizens'	communality, collectivism, egalitarianism	city as a communal infrastructure	centripetal →static
Alternative state-socialist counter-public	mid-1950s to 1990s	'we, the people'	collective solidarity	city as a space for contestation	centripetal impulses
Post-socialist public sphere	1990s to the present	atomistic individuals	entrepreneurship, privatism, (sacred) right to private property	city as a commodity	centrifugal
Post-socialist counter-public	2010s to the present	urban grassroots movements, new bourgeoisie	right to the city, concrete narrative, spatial justice	city as a commons	centripetal

around 2007, with Romania and Bulgaria joining the EU in the second wave of accession in the region, three years after the Baltic states, the Czech Republic, Hungary, Poland, Slovakia, and Slovenia. During the next, third decade of 'actual' post-socialism, local counterpublics emerge, with a competing programme and urban paradigm.

Urban common good under socialism

Whether established before or built from scratch after WWII, all cities in CEE underwent the same ideological (re)modelling after 1945. In essence, it boiled down to the communalisation of urban resources, collectivisation of urban living, as well as application of egalitarian principles in urban planning and other urban policies (Enyedi 1996, de Magistris 2012: 31). Public discourse was dominated by catchphrases and slogans such as the one pronouncing Warsaw 'the socialist capital city for every citizen: worker, peasant and intellectual' (Crowley 2002: 191). Presenting its six-year reconstruction plan in 1949, Bolesław Bierut, President of Poland and General Secretary of the Polish United Workers' Party, vowed that the rebuilt city would not become 'only an improved duplicate of the pre-war assemblage of private interests within the capitalist society' (Bierut 1951: 121).[3] Elimination of private property and differences in social status resulting from gender inequality, establishment of a classless society, and empowerment of the proletariat were to restore social justice absent in bourgeois democracy (Lenin 2001). Unlike capitalism, 'based upon instrumentality, war and competition' (Stevenson 2017: 18), socialism was to rely on improving the quality of life in a non-exclusionary, egalitarian way.

Ideologically robust and convenient, social justice would then become an umbrella concept setting the dominant discourse in the official public sphere and providing an overarching aim for the principle of urban common good. Back in 1872, Friedrich Engels wrote that 'it is not the solution of the housing question which simultaneously solves the social question, but only by the solution of the social question, that is, by the abolition of the capitalist mode of production, is the solution of the housing question made possible' (Engels 1887). Elizabeth Zarecor (2018: 106) reads this quote as not just critical of handling the housing crisis solely through housing provision, but also as a call for a 'complete reorganization of everyday life toward communal living'. Acknowledging the historical effects of this appeal, which materialised in the Eastern Bloc around the 1940s, Zarecor (2013, 2018) proposes to analyse the socialist city using two complementary frameworks: infrastructural thinking and the socialist scaffold. She defines the former as 'decision making propelled by the requirements and scale of urban infrastructure' (2018: [6]). The latter—consisting of such interrelated elements as public transportation systems, housing complexes, and green and recreational spaces, designed to provide for and support the centrally planned economy and the collective way of life—adds up to the urban infrastructure itself.

Investigating cities through their infrastructures in a broad sense of the word, involving not just technicalities but also social and political properties, provides substantial, multidimensional information on the urban condition. It is also useful as a conceptual frame of reference for looking at a city as a collective resource:

> The design politics of infrastructure is a blueprint of how the city is imagined and constituted as a collective surface, as a shared life-support system, or as a field of scarce and rival resources. The stuff of infrastructure keeps life together in the big city at the same time as it works to hold people, spaces, and resources apart.
>
> (Tonkiss 2013: 156–157)

As much as this approach remains universal, it seems to be particularly suited for CEE cities during most of the second half of the 20th century, when the construction of new and modernisation of old infrastructures for the glory of socialism monopolised the narrative of urban development (Tuvikene et al. 2019). Therefore, the state socialist model of urban common good leans on the paradigm of the **city as communal infrastructure**, marking the rejection of the capitalist vision of the city as a space for consumption.

In this scenario, the city as such was 'unimportant'; it merely served as the backdrop for the formation of the state socialist order. Its role was reduced to a vehicle for change, necessary to set the project in motion. This was augmented by distorted property relations. The fact that most Eastern bloc countries disqualified private property, allowing—apart from personal possessions—only social ownership (i.e., by the state, a cooperative, or other collectives) (Marcuse 1996), bred the paradox of no one's property: the perception of anything common as 'belong[ing] to all and to none' (Kornai 1992: 75). Political sociologist and economist Jacek Tarkowski attributes this to the process of depersonalisation of collective property, leading to a declining sense of responsibility:

> Abstract and ill-defined common property tends to be treated as no one's property, or rather ceases to be treated as property at all. On the one hand, it escalates the sense of alienation, that is, the sense that one neither controls the reality, nor is the means of control clearly determined, and on the other—[it amplifies] the sense that 'common' signifies 'ownerless'.
>
> (1994: 275)

In practical terms, the logic of no one's property legitimised putting individual interests before the common good, whether it concerned stealing office supplies from the workplace or vandalising public space, even under threat of penalty.[4]

The role of the citizens as collective urban actors was also perfunctory. What follows from the totalitarian nature of the official public sphere in CEE during state socialism is that they were politically redundant. Deprivation of liberal democratic rights rendered them 'citizens' solely by name. Not only were they disempowered, or even unable to voice their concerns, but their freedoms were also restricted through coercive control of the state and the socialist community:

> In the people's democracies the idea of the common good, which was the objective of the public sphere, went far beyond the traditional notion of seeking consensus about what is best for everybody. Within the common sphere, the criteria for personal well-being were determined by the community according to its own criteria, based on the personal convictions of its members rather than the wants or needs of the individual concerned. The public space was engulfed and smothered by the collective. People found themselves in an inferior position in relation not only to the State but also to the socialist community, which demanded that individuals must conform to its expectations.
>
> (Krakovský 2018: 101)

To sum up, the model of urban common good circulating in the official public sphere under state socialism comprises three areas which are adjacent but not intersecting. Passive 'citizens' (collective human actors) inhabiting the city reduced to communal infrastructure (urban paradigm) are held back from collective action, regardless of the superficial doctrines of communality, collectivism and egalitarianism (dominant rationales). Since all three elements are imposed top-down, they do not provide grounds for bottom-up production of the common good. The artificial category of 'citizens', who, instead of being actively engaged in the creation of the socialist urban space, perform stand-in roles for the actual decision-makers, is just as instrumentalised as the idea of social justice that legitimises the authoritarian regime.[5] The common good thus remained deadlocked until the rise of the alternative public sphere, which prompted a game-changing reconfiguration of the model.

Although theoretically consistent with the principle of social justice, the concept of city as communal infrastructure failed to work in practice shortly after it was introduced. Its promise of centripetal energy lost momentum as inequalities and inefficiencies of the communist system—particularly those connected to the provision of infrastructure—led to the dismantling of 'shared spaces of common life' into 'zones of differential access and exclusion' (Tonkiss 2013: 173). The ensuing disillusionment is partly responsible for triggering social unrest, which eventually contributed to the fall of communism across CEE. Even if the alternative public sphere did not specifically aim to address urban issues, it was definitely anchored in urban space, where counteracting the official public sphere actually *took place* through collective action based on civic consolidation and solidarity (Lang 2013). The sets that make up the model of urban common good in the alternative counterpublic includes 'the people' (collective human actors), mobilised by the idea of collective solidarity (dominant rationale) and appropriating the **city as a space for contestation** (city as a collective resource). Unlike in the static model of urban common good characteristic of the official public sphere, here all three components are bound together by a centripetal force that taps into the potential of urban commonality.

The outburst of long-cumulated bottom-up energy, manifested in the revolutions of 1989, led to high expectations towards the young democratic societies. Yet the vision of 'a golden age of West European-style democracies, renewed public realms and widespread civic engagement' soon faded, as by the end of the 1990s, the same citizens, collectively 'retreated from the public sphere' and 'rushed toward purely private concerns' (Hirt 2012: 22). Therefore, the moment when the communist domino stones started to collapse, country after country, is where the plot twists once again.

Urban common good during transition

The caesura of 1989 theoretically closes the era in which CEE citizens were able to negotiate the common good only outside the official public sphere, having limited capacity to act openly in urban space and nothing but the clandestine media at their disposal. On the one hand, the alternative sphere could take it from here and naturally morph into a genuinely public one. On the other—the socialist implosion culminating in the 1980s had gradually instilled a certain aversion to anything common or communal, reaching beyond the disregard for no one's property. Before the last of the state-socialist regimes collapsed, it was clear that the bottom-up energy was used up and the alternative public sphere would not persist in the same shape under the new conditions.

Moreover, the long-awaited capitalist order heralded individualism, entrepreneurship, and competition as the core cultural values. It fed on anti-socialist sentiments, deeming 'the common' an outdated product and a rotten symbol of the bygone era. Such treatment incited 'scepticism towards the social' (Sidorenko 2008: 110) and enabled utilisation of the recent past as the '"ultimate bogey" for pre-empting social claims' (Chelcea and Druţă 2016: 525). Elizabeth Dunn's (2004) case study of privatisation processes in a baby food plant located in a medium-sized Polish city of Rzeszów reveals how this philosophy engineered the breakdown of social ties in formerly state-owned workplaces—wellsprings of Solidarity only moments before.

Consequently, in the shift from planned to free-market economy, 'citizens' and 'the people' alike were replaced by atomistic individuals uninterested in performing together for the sake of the common good. Not only the public now ranked much lower than the private—a globally upward trend since the 1980s—but almost all efforts towards collective action were abandoned, abruptly and categorically (Cymbrowski 2017: 18). The urban arena became increasingly divided between conflicting categories of actors in varying positions of power—tourists versus inhabitants, inhabitants versus entrepreneurs, drivers versus pedestrians, pedestrians versus cyclists, and so on (Sagan 2000).

These tendencies were reinforced by the rehabilitation of private property rights, which became constitutionally guaranteed in all countries across the region. Legal provisions involved discarding 'social' and acknowledging 'private' ownership; they secured rights for the latter but generally maintained, although to varying degrees, public priority in the rights to land and natural resources (Marcuse 1996: 171). The 'sacred right to property' soon rose above public interests and trumped them (Stanilov 2007, Drozda 2017), becoming endorsed as the prevailing discourse of the new official post-socialist public sphere. Property relations, emerging as the 'fundamental force behind the transformation of the use of urban space' (Bodnár 2001: 11), contributed to throwing apart the elements of the urban common good and loosening the relations between them. Its centrifugal dynamic supplanted both the inertia of the state socialist public sphere and the centripetal motion of the alternative counterpublic.

The welcoming of the market as a key regulator and driving force of urban change made the establishment of the paradigm of **city as a commodity**, borrowed from the West, all the easier. Much of what used to be common urban infrastructure underwent fragmentation, commercialisation, and appropriation. Trends such as residential gating, expansion of shopping centres, and gentrification have cemented the new order introduced along with the free-market economy. Cultural landscapes of post-socialist cities relieved from the socialist propaganda of the communal and (pseudo)common have surrendered to individualism, privatism and the imperative of profitability (Czepczyński 2008).

But while transformations in the post-socialist public sphere and public space emulated those in the neoliberal West, they were also tailored to post-socialist conditions. To begin with, the withdrawal of the state was more acute, as decentralisation burdened local governments with numerous obligations and insufficient funding. The combination of task overload and immense budgetary pressures considerably hindered the effectiveness of municipalities. This partly explains the subsequent retreat of local policy-making, especially in terms of urban planning and spatial development. Furthermore, public regulation and control remained influenced by the relics of the previous regime. This could be observed, for instance, in the tendency of the new power coalitions to step into the shoes of the old elites. As observed by Zarecor (2018: 112), in Czech and Slovak cities, 'an emerging network of real estate developers, international banking managers, and local corporate executives and politicians (some of whom came to their wealth under suspicious circumstances during privatization) have replaced the monolithic group of Communist Party apparatchiks'. Lastly, the return to liberal democracy did not automatically translate into citizen participation in decision-making at the local level. While more and more Central and Eastern Europeans chose to settle in urban arrangements, they were not encouraged to look beyond their particular interests or the restrictive standards of representative democracy.

In the post-socialist city at the turn of the 21st century, urban common good drowned in a sea of individual interests, among which those of the citizens were the most underrepresented. Although public discontent was less pronounced than two decades before, it brought about a proliferation of urban movements, which by the end of the 2000s effectively questioned the excessive privatisation and commodification of the city within the new order. They soon gathered enough impetus to develop a distinct counterpublics—just as predicted in the second scenario by Jakubowicz.

Urban common good in post-socialism

What made the mobilisation of urban movements in CEE distinct in comparison with the West is that it occurred in the absence of fully fledged civil societies. Judging by Western standards, that is, merely by the number of professional NGOs or by formal civic engagement, they were relatively weak indeed. Following their Western counterparts in channelling dissent against neoliberalism, they usually took off as small-scale, informal initiatives, only some of them working their way to more institutionalised forms and modes of operation (Bitusíková 2015, Pixová 2020, Domaradzka 2021). Kerstin Jacobsson (2015) claims that their grassroots nature postponed recognition of their potential, which explains why around the 2010s they took the urban stage in countries across the region as if by surprise.

Jacobsson's perspective spotlights the shifting identity of urban dwellers: from atomistic individuals pursuing their own interests towards not just collective but collaborating actors. Michaela Pixová (2018: 673) attributes the origins of Czech urban grassroots initiatives to the citizens' 'increasing emancipation, education, and awareness in relation to the city and urban processes', along with their growing demands 'as recipients of urban services, politics, and the quality of urban environments'. The formation of a novel type of citizen is also observed by Paweł Kubicki (2011: 226), who identifies the new bourgeoisie as locally engaged 'individuals in action using private resources to achieve public goals'. Resources here refer to the social and cultural rather than financial capital. The term clearly alludes to the Habermasian public, but Kubicki leaves the category broad, comprising a mix of features representative of socio-cultural urban groups known from the West—the 19th-century flâneurs, the late 20th-century yuppies and bobos, the millennial members of the creative class, and the more contemporary hipsters.

The rediscovery of urbanity and urban commonality in CEE has been paired with a recognition of the right to the city. Adopted as a guiding framework, it affiliates the postulate of a participatory approach to decision-making as well. Its acknowledgement is evident in the names of many contemporary urban movements, such as *Miasto Jest Nasze* (The City is Ours) in Warsaw, or *Praha sobě* (Prague for Us). In Poland, the footing of the right to the city is further supported by the rationale of 'concrete narrative' (*narracja konkretna*), publicised by Poznań-based urban activists – Lech Mergler, Kacper Pobłocki and Maciej Wudarski (2013). They who claim that collective action and building coalitions across different citizen groups for the sake of urban common good is more effective when they rise above typical political divisions and focus on concrete issues to be solved. In this respect, the concept is very much in line with John Ryder's pragmatic approach to politics (see Chapter 1).

The remaining third element of the model of common good applicable to the post-socialist counterpublic is the paradigm of the **city as a commons**. However, as outlined in Chapter 1, the concept of commoning in CEE bears a specific legacy. On the one hand, it draws from the socialist tradition, hoping for a return of the city as communal infrastructure, although this time with a more reliable mechanism of spatial justice built in. On the other hand, it turns away from this very tradition, by demanding citizen participation in governance. Both conditions make the paradigm not entirely congruous with its understanding in Western Europe.

As evidenced, transformations of the model of urban common good during the transition and in post-socialism strongly resemble Western European patterns in some respects but diverge from them in others. In terms of similarities, both sequences feature a corresponding rhythm of ebbs and flows of the collectivity factor, with intermittent reorientations of the model's driving forces—from centrifugal to centripetal and back again. In both cases, centripetality tends to surface in the counterpublics, by way of opposition to the discourse established in the mainstream public sphere. Moreover, what is characteristic of both strands is that the weight of the urban paradigm builds up over time. Previously reduced to spaces of debate or consumption, cities on either side of the former Iron Curtain with time became valuable commodities and valued commons. A gradual change in the politics of scale is visible here, namely urban emancipation from state ideology and legislation at the national level. Finally, independently of the particular policies of individual countries, global trends have increasingly affected the local (Sýkora 2008), being responsible for the apparent, although incomplete, convergence of the two trajectories.

Equally important are the discrepancies between the two threads of development, related to the path dependency of the post-socialist urban common good reaching back long before the end of WWII. Although pseudopublicness of the early capitalist period had given in to the state-imposed socialist public sphere, which strongly emphasised urban common good, within only a few decades the initially centripetal configuration of the model came to a standstill. This, however, did not prevent the idea of city as communal infrastructure from becoming culturally imprinted into the societies of the Eastern Bloc. In simplified terms, urban common good became an issue in the West due to the emergence of a bourgeois public sphere, quite contrary to the Central East, where it developed into a public priority owing to the state socialist ideology.

Nonetheless, the subsequent discrediting of the socialist city's moral values gave rise to a counterpublic which, guided by the principle of collective solidarity, aimed to continue the unfulfilled promise of social justice (cf. Smith 1994), reapplying centripetal acceleration to the model. Consequently, its capacity was much larger than in the analogous alternative public sphere in the West. With the advent of 1989, the model's dynamic changed once more. Approximating the neoliberal public sphere in the West of Europe, the urban discourse in post-socialism went much further in rejecting urban commonality. The anti-collective CEE logic of the sacred right to property eclipsed Western European tendencies of individualism and privatism. However, the quickly progressing commodification of the post-socialist city set the agenda for another counterpublic, whose centripetal potential yet again outperformed the Western one. Remarkably, the released social energy of the new bourgeoisie, complementing that of the new urban movements, to a certain extent parallels the Western bourgeois revolution from two centuries ago.

The remainder of the chapter is organised around urban paradigms characteristic of the three consecutive periods. It is through the lenses of the city as a communal infrastructure, city as a commodity and city as a commons that I will investigate the transformation of the CEE models of urban common good in closer detail, capturing the specifics of the changes in selected areas that critically influenced the making of urban common good under socialism, during the transition, and in post-socialism.

City as a communal infrastructure: the rise and demise of the socialist urban utopia

From the today's perspective, the imposition of communist ideology in 20th-century Central and Eastern Europe was doomed to fail, and it is taken for granted that this failure accounts for the discredit of communal values after 1989. To evade the trap of presentism, however, it is important to bear in mind that despite their brutality and oppression, authoritarian state socialist regimes were initially quite widely supported. Exhausted and traumatised in the aftermath of the WWII, many citizens of the newly established people's republics banked on rebuilding their countries in a different, more equitable way, choosing to believe in the communist utopia (Zaremba 2012). By the end of the 1940s, Polish workers aspiring to a better future,

> strived, not always consciously, to bring into being a Poland that seemed fair to them, namely, without poverty, without exploitation, without bourgeoisie and without strangers—in which they would decide themselves about "their" workplaces and be able to choose employment on their own accord.
>
> (Kenney 2015: 16)

It is widely documented that towards the end of the war, employees of industrial plants in larger cities hid vital parts of machinery to prevent them from destruction by the retreating German troops. This strategic appropriation of industrial infrastructure may be regarded as a specific case of urban commoning: 'In the spring [of 1945], they reassembled [the machinery], restarted production and considered themselves the rightful owners of the city' (Pobłocki 2011: 138).

In CEE, the need to rebuild war-torn cities was urgent and apparent, and it also legitimised the ideology of the new order. After all, communism endorsed working together for a common cause, and this particular exigency required no further substantiation. Recounting the zeal of the very first reconstruction efforts in Warsaw, Grzegorz Piątek (2020: 38) mentions the blurred boundaries between spontaneous collective action and top-down initiative. He enumerates how schools were improvised in private homes, firemen raced on foot to get to emergencies, power engineers from all over the city banded together to restore electricity, and tram drivers assembled replacement vehicles from the broken remnants. Civilians actively participated in the restoration of the ruined capital, but similar fervour was observed in second-order urban centres. Subbotniks, voluntary work assignments performed on Saturdays, were massively attended even in the former Prussian cities, culturally foreign to the newcomers from the East, such as Gdańsk or Wrocław (Perkowski 2013, Kenney 2015).[6]

But aside from the short-term objective of post-war recuperation, the communist project was aimed at modernisation through forced industrialisation. The latter quickly became a top urban priority (Gentile and Sjöberg 2006), and the development of urban infrastructure—arteries complementing the bone structure of the urban form (Tonkiss 2013: 138)—was central to this process. Popular songs of the 1950s presented the city as a festive construction site, evoking an atmosphere of eagerness, optimism, and pride. In the Polish hit composed by Władysław Szpilman, a tellingly red bus rushes 'through forests of scaffolding', 'past the new, bright houses' to deliver commuters to their workplaces on time (Winkler 1952). As observed by Tuvikene et al. (2019: 10), even though urban infrastructure 'remained auxiliary to industrialization rather than a value in itself', it brought tangible benefits to the inhabitants:

> combined with mass housing construction, literacy campaigns and health care expansion, major sectors of the population profited from Soviet infrastructure investments. Based on the socialist ideology of equality, these included an exclusively state-sponsored provision of services and particular cost-calculation regimes. This led to less polarized spatial coverage of services, and relatively equal access to diverse forms of infrastructure such as transport, housing-related infrastructures, and urban greenery.

However, all these reconfigurations, seemingly accomplishing the ideal urban environment for an egalitarian collective society, came at many costs. One of them was land nationalisation—a prerequisite for the paradigm of communal infrastructure to take hold, as it 'meant the possibility of a new spaciousness, a completeness, the treatment of any given site as a potentially blank slate, and the possibility that town plans could be completed without the obstruction of any private interests' (Hatherley 2015: 10). Hence, the installation of the infrastructural order demanded that citizens surrender the right of ownership in the city but did not offer them the right to the city in return. Regardless of their involvement in its construction, they remained voiceless users with little say on the communal infrastructure—the handover was purely ideological.[7]

Communal infrastructure is an all-encompassing term. In the strict technical sense, it combines the interlinked categories of energy and electricity networks, water supply and sewage systems, public transport, roads and road facilities, urban greenery, and solid waste management. However, the list can be extended to include social amenities and services, such as public housing and public health programmes. For the purposes of this chapter, I will focus on four key issues which constitute four critical spheres of urban commonality, i.e., housing, transport, green infrastructures, and public space. Apart from providing solid foundations of urban hardware (Tuvikene et al. 2019: 6), they were crucial in the formation of a new collective urban identity in CEE after 1945. The development of municipal housing, collective transport, and green spaces occupied a prominent position in the official public sphere, and any successes in these fields generously fuelled the state propaganda machine. Public spaces, accounting for both the hardware and software of socialist urban commonality, performed the role of social condensers—shaping social relations by design (Kopp 1970). Taken together, transformations within these four areas of communal infrastructure recount the story of how the socialist city came together to fall apart.

Housing

Most cities in CEE were affected by shortages of dwellings even before 1939. Due to war damage and massive immigration, intensified by the post-war baby boom, the housing crisis in the emerging socialist regimes reached an unprecedented scale, much surpassing the pre-war deficits. The provision of urban accommodation thus became a pressing issue, entirely incumbent on the hegemonic state. One immediate solution consisted in the redistribution of the existing dwellings through the Soviet institution of *kommunalka*, in which apartments were shared between several households.[8] Meanwhile, the progressing industrialisation necessitated extensive construction of housing estates for workers. The final years of Stalinism saw large residential projects of this type. In Warsaw, the openly revanchist vision of 'people entering the city centre' materialised in stone in 1952, with the completion of the Marszałkowska Residential District (MDM)—a monumental complex erected in the war-torn southern part of the city centre. Built in the characteristic socialist classicist style, expressed both in the urban layout and architectural details, it echoed similar undertakings in central and inner neighbourhoods of other CEE cities. Such blueprint developments, conforming to the design paradigm of 'national in form, socialist in content', include the Stalinallee apartment blocks in East Berlin and The Largo ensemble in Sofia.

Socialist housing utopias also emerged as entirely new cities, either showcasing heavy-industry establishments (Stalinstadt/Eisenhüttenstadt, Dunaújváros, Nowa Huta) or lodging the industrial workforce in large-scale residential estates (Poruba). Some of them, located in the vicinity of old-bourgeois urban entities to serve as their proletarian counterpoints, eventually merged with them. Nowa Huta (New Steelworks)—with the Vladimir Lenin Steelworks employing almost 40,000 people at its production peak in the 1970s (Lebow 2013)—became a district of Kraków in 1951, while Poruba—a monumental worker suburb 'advertised as the creation of a beautiful place for miners to live' (Zarecor 2011)—was incorporated into Ostrava in 1957.

Since the grand projects of the Stalin era were insufficient to meet the increasing demand for housing infrastructure, socialist states redirected investment to the construction of large, prefabricated housing estates—finally an efficient response to the incessant flow of migrants from rural areas (Fidelis and Gigova 2017). In the 1960s and 1970s, CEE cityscapes became irreversibly imprinted with mass-produced, uniform slabs and tower blocks. By realising the vision of social egalitarianism, both the excessive standardisation and collectivity of these new housing units fulfilled the ideological requirement of mixing social classes (Kovács and Herfert 2012, Szafrańska 2015). The flats may have been small, but the emphasis was put on common facilities, intended to promote communal living. The urban design, packed with social infrastructure—pre-school and educational facilities, primary healthcare centres, shops, and playgrounds—was intended to address individual needs and to foster collective interaction, turning residential complexes into social condensers. Much consideration was given to planning at the level of self-contained residential neighbourhoods with basic public services and facilities emulating the Soviet microdistricts (microraions).

For many families moving into large housing estates, this meant a considerable improvement in quality of life. At the same time, it is debatable whether the model of forced community-building through spatial planning and residential design served the urban common good. Václav Havel claims otherwise, insisting that the seemingly communal qualities of the prefabricated housing estates diverted focus from the hidden process of reorientation of social energy. Switching from the 'outward' to the 'inward' mode, in 'an escape from the sphere of public activity', people would turn 'their main attention to the material aspects of their private lives' (Havel 1990: 11–12).[9] Consequently, the barely launched process of the formation of collective urban identity was hampered by the depoliticisation of the public sphere and withdrawal into self and individual aspirations.

Moreover, the anticipated integrating effects of socialist housing schemes were called into question by the shortcomings of the centrally planned economy, manifest in underinvestment in basic urban amenities. Provisional pavements, muddy to-be lawns, and unfinished retail premises all became symbols of urbanisation without modernisation (Węgleński 1992). This contributed to the perception of common space as no-one's—collective, but at the same time amorphous, badly managed, and used irresponsibly (Królikowski 2020: 93).

As a result, the city and urban space became increasingly perceived as belonging to nobody. By way of irony, in Czechoslovakia and Poland, a significant proportion of such 'ownerless' spaces was run by housing cooperatives. By the end of the 1980s, this type of tenure, classified as a form of social property, amounted there to around 20 and 25 per cent of dwellings, respectively (Struyk 1996). Due to corruptive practices and the general inefficiency of the centralised system planning and control, their potential for collective action remained untapped (Matysek-Imielińska 2020).

Lastly, despite the official egalitarian doctrine, the access to accommodation was unequal. Due to corruption and the renewed problem of housing shortages, the constitutional right to housing and the predominantly public ownership of residential property did not guarantee a roof over one's head. Especially towards the end of the 20th century, the 'better connected' members of society enjoyed more housing opportunities than the common people (Bodnár and Böröcz 1998). Moreover, contrary to the state propaganda, ethnic segregation and ethnic inequalities, which were to vanish during socialism, persisted in this domain (Gentile and Sjöberg 2006: 709).

Public transport

The availability of public transport was essential in rapidly growing socialist cities, where the levels of car ownership remained relatively low (Siegelbaum 2011). Dense networks of bus, trolleybus, and tramway routes developed at an accelerated rate, providing efficient and fairly reliable daily commuting options. Almost forty sections of tramway and trolleybus lines were built between 1950 and 1989 in Brno (Mulíček and Seidenglanz 2019). In the mid-1980s Budapest, public transport covered 61 per cent of daily traffic entering the city (Machon and Dingsdale 1989).

This rapid expansion of public transport required immense investment efforts. New roads and railways, tunnels, and bridges emerged as costly, and at times spectacular, centrepieces of socialist urban planning. Public media closely reported the progress on any larger projects; their completion was celebrated with functions featuring official addresses by communist party leaders. Citizens actively participated in the construction works. A quasi-voluntary engagement of Praguians contributed to the development of tramway tracks in the capital of Czechoslovakia (Žídek 2019). In Warsaw, the effort put into the W-Z Route, a communication axis bringing together the city on both sides of the river Vistula, was pompously compared by the contemporary writer Jarosław Iwaszkiewicz to the 'collective expression' of the ancient Greek theatre and mediaeval cathedrals (Piątek 2020: 338) (Figure 2.2).

Modern transport infrastructure helped to perpetuate the myth of the socialist city as well-organised, sustainable, and socially just (Czepczyński 2008: 66). One of the most notable symbols of urban prosperity in Central and Eastern Europe was the underground. Its cutting-edge technology and aesthetic design—especially lavish in Soviet Russia—heralded social progress

Figure 2.2 City as a communal infrastructure: entrance to the construction site of the W-Z Route in Warsaw. The information board reads: 'The W-Z Route is co-financed by the Social Fund for the Rebuilding of the Capital'.

and an improved quality of life (Hatherley 2015), while also fulfilling the principles of egalitarianism and communality. Since the 1940s, over a dozen underground networks were built in the USSR or exported to the capitals of its satellite states. In some cities across the region, underground systems had been planned before then, but the construction was postponed because of the outbreak of WWII. Their post-war expansion was hence promoted as a reproduction of the Soviet technical thought.[10] Regardless of tremendous delays in many of the realisations—in Warsaw the first line opened in 1995, following 12 years of construction, and in Sofia, the inauguration took place in 1998, over 30 years after the plans had been made—they were deemed remarkable achievements and an ideological success.

Available only to inhabitants of larger cities, the underground was the epitome of collective urban transport. However, the automobility project was not given up. Eastern Bloc urban planners of the 1960s and 1970s acknowledged the need for individual transport and modern road infrastructure, and went on to redesign cities accordingly, introducing multi-lane arteries, flyovers and tunnels. Exaggerated at the time, these infrastructural interventions found application only after 1989. At the end of the 1980s, the maximum waiting time for a car—an increasingly valued object of desire—still ranged from two to three years in Czechoslovakia to seventeen in East Germany (Kornai 1992: 236).

Green infrastructure

To fully appreciate the significance of green infrastructure in the socialist city, one should think of the pre-war cities in CEE, densely developed and lacking open green areas. Deemed unprofitable, such use, or rather 'waste', of space was incompatible with the capitalist agenda. In contrast, the multiple benefits of urban greenery became greatly appreciated after WWII. Although the imperative of industrial development in the Eastern Bloc generally worked against natural environment, urban green infrastructure was a notable exception to the rule. Socialist urban planning guaranteed proximity to nature to all citizens (Gibas and Boumová 2020), and—just like in the case of housing and transport infrastructures—nationalisation of land and absence of the bid rent mechanism accommodated the expansion of greenery at a scale unattainable in old-capitalist democracies.

The components of urban green infrastructure ranged from public parks, boulevards, and promenades, through botanical gardens and allotments, to horticultural, orchard, and flower farms. They were aimed at separating functional zones and served multiple purposes—recreational, social, aesthetic, and economic. Most of them were available for general use, while access to some was more restricted, yet all of them were subjected to rigorous planning regulations. For instance, a set of Polish guidelines from 1974 determined such details as the minimum hours of sunlight in playgrounds (Oględzka 2010: 271). Special attention was paid to ensure access to open green spaces in the newly constructed large prefabricated block estates. According to the guidelines of the Athens Charter, the provision of 'massive amounts of public greenery between buildings' was a must (Hirt 2007: 153), especially that the inhabitants were deprived of individual gardens.

Partly compensating for this lack, vast recreational grounds, modelled after Moscow's Gorky Park, flourished in the first- and second-order cities. In Romania, for example, between 1948 and 1990, the number of urban parks increased tenfold, and their total surface tripled (Badiu et al. 2019). Unlike 19th-century pocket parks—fenced, guarded, and frequented mostly by better-off city dwellers—their oversized, socialist versions were open for everyone. Combining scenic walking paths and ponds with sports facilities and children's playscapes, they admitted visitors of all ages and backgrounds either free of charge or at a nominal fee. The largest Polish complex of this kind, the General Jerzy Ziętek Voivodeship Park of Culture and Recreation, opened in Chorzów in the 1950s, was hailed as the 'green lungs of Upper Silesia'. Covering over 600 hectares of former mining wastelands in the most heavily industrialised and densely populated region in Poland, it offered a long list of attractions and facilities, including a planetarium, a football stadium, a fairground, a zoo, and an aerial tramway.

Another type of collective green spaces typical of socialist urban landscapes was allotment gardens. After WWII, they were already present in cities across the region, but were especially promoted for similar reasons as recreational grounds. Other than that, they functioned as an additional source of food supply—a significant role in the context of perpetual food insecurity (Bellows 2004).

Regardless of the attention paid to urban green infrastructure by the socialist authorities, in many locations it underwent considerable degradation over the years. Common green spaces between residential blocks, parks, and other common grounds dilapidated due to insufficient funds and poor management (Brzostek 2007). This, apart from transnational influences (Carmin and Hicks 2002), contributed to the rise of environmental activists, who had their small share in the later demise of the socialist regimes in CEE.

Public space

Before WWII, public spaces of CEE cities were dominated by private interests similarly to their Western counterparts. One of the roles of urban planners after 1945 was to reclaim these spaces by overcoming their stigma of elitism. To this end, the bourgeois veneer of central districts was disposed of through functional transformations. Working-class accommodation claimed the territory formerly reserved for elegant boutiques, bistros, and cafés. Invalidation of the bid rent mechanism effectively minimised commercial use of urban space in previously prestigious locations.

In addition to being egalitarian and evenly distributed across the city, the new public spaces acted as social condensers. Instead of perpetuating disintegration, they were to 'stimulate the pulse of social life' and accommodate the 'river of collective coexistence flowing through the city' (Goldzamt 1971: 257). This broadened the concept of public space to include not only open-air recreational grounds, but also libraries, cafeterias, and clubs. This vision found its perhaps most prominent realisation in the palaces of culture. Intended for cultural education and leisure for the masses, they were usually quite impressive edifices, housing cinemas, theatres, conference halls, exhibition spaces, libraries, and museums. The iconic features of the 1950s Palace of Culture and Science in Warsaw, or its younger siblings completed in Prague and Sofia at the beginning of the 1980s, not only represented the official culture, but also accentuated its liberation from the former establishment, with a symbolic reconfiguration of power relations indicated in the name (Murawski 2019).

It would thus seem that in many CEE cities under socialism 'the people' indeed 'entered' the centre. However, even if the accessibility and inclusiveness of public spaces improved, this came at the cost of their publicness, which was controlled top-down, both in terms of planning and use. On the one hand, urban streets and squares served to legitimise the socio-political order by hosting the 1st of May or military parades. On the other, apart from the rare moments of regulated celebration, they were rather de-publicised and de-politicised. Appropriated by the regime, devoid of authenticity and closed off to any bottom-up initiative, public spaces became vacuous and sterile. A notorious example of their instrumentalisation is the 1980s construction of the monumental Centrul Civic in Bucharest, which required the erasure of much of the historical urban fabric in the heart of the city. Similarly destitute and declarative were other sites of power in socialist cities, such as Plac Centralny in Nowa Huta or Alexanderplatz in East Berlin. According to Czepczyński (2008), public spaces of this kind turned into anti-agoras, where the only permitted acts of civic engagement were manifestation of obedience and support for the ruling party. Only at times of upheaval, when counterpublics would momentarily emerge, the anti-agoras resumed their role as genuine public spaces, where citizens could voice their discontent.

A specific type of (counter)public space emerged in Polish and East German cities, where religious practices flourished despite the strictly secular programme of the socialist state. With the development of the alternative public sphere, church buildings became more than houses of worship. Bringing together local communities, they also sheltered oppositionists and served as distribution points for foreign charity aid (Cordell 1990, Nawratek 2005). New temples mushroomed in the quickly urbanising Poland owing to the negotiations, periodically faltering, between the Roman Catholic hierarchs and the communist government (Kucza-Kuczyński and Mroczek 1991). Left out of control by the communist system, the edifices were typically designed, built, and co-funded by self-organising citizens (Cymer 2019). Sacral architecture—also dubbed Day-VII Architecture or architecture of protest (Cichońska et al. 2019)—often acquired a political edge. A compelling example comes from Nowa Huta, where in 1960 the Party's withdrawal of its initial permission to erect the church of Our Lady Queen of Poland mobilised collective action of the local citizens. Following a wave of protests, the building was completed a few years later.[11]

This single act of civil disobedience was one of the many preludes to more concerted actions which would take place throughout CEE just a few decades later, during the revolutions of the 1980s and at the turn of the 1990s. In August 1980, crowds gathered at the gate of the Gdańsk Shipyard to show support for striking workers. In 1989, Budapesters participated in a symbolic burial of Imre Nagyi in Heroes' Square, Leipzigers chanted 'we are the people' in Karl-Marx-Platz during Monday Demonstrations, Berliners celebrated the fall of the Wall at the Brandenburg Gate, and velvet revolutionists took to the streets of Prague. Over the next two years, they were joined by protesters in Timişoara, Bucharest, and other Romanian cities. As noted by Stanilov (2007: 269), '[f]rom symbols of totalitarian oppression during the second half of the twentieth century, the main public spaces of Eastern European cities turned into a dramatic and potent stage [of] the heroic struggle for democracy'. In this sense, public space of socialist cities finally lived up to the name of communal infrastructure.

City as a commodity: privatisation and appropriation of the common since 1989

The transformation of the urban paradigm in CEE after 1989 may be investigated in terms of continuity versus change within the framework proposed by Tauri Tuvikene (2016) in his concept of deterritorialised post-socialism (see Chapter 1). In this context, continuity is often understood as the persistence of undesirable arrangements and patterns existing in the former system. For instance, in post-socialist urban planning, it is equated with the durability of the silo approach and insufficient horizontal coordination (Nedović-Budić 2001). But the socialist legacy at the turn of the 1990s also had positive potential—for example, the social mix, well-developed networks of public transport and green infrastructure, and compactness. Zorica Nedović-Budić (2001: 47, drawing on Maier 1998) attributes these advantages to the relative 'backwardness' of CEE cities, but also asserts that 'good examples of planning as practiced during the communist regime are easily forgotten'. Similarly, Elizabeth Dunn (2004: 164) notes in her book on privatisation in the Polish industrial sector that 'in eradicating the institutions of state socialism, neoliberalism and post-Fordism have damaged or destroyed some of the most socially valuable aspects of the socialist era'.

Across the region, the transition period was therefore marked by a rather radical urban change. The embracement of neoliberalism, facilitated by a strong belief in the reliability of market mechanisms and the need for privatisation of public services, turned to a much quicker and more aggressive adoption of the doctrine than in the West. Moreover, even though in most countries decentralisation processes enabled emancipation and empowerment of local communities, the opportunity was not seized immediately. Due to legacies of the past (the strained condition of social capital, measured by low levels of trust and cooperation) and changing conditions of the present (the initial hardships of soaring unemployment and inflation, and the subsequent reorientation towards privatism and consumerism), urban dwellers once again abandoned the public sphere for over two decades. Even the oppositional church-aligned domain of resistance lost its grounds.

The resulting 'galloping commodification' (Tonkiss 2013: 143) of the post-socialist city translated into its increasing reliance on private capital, focus on profitable megaprojects, and place branding (Szmytkowska 2017). Despite the overly promoted narrative of sustainable development, it was the economic aspect that gained the most prominence. Both social and environmental concerns were downplayed on the assumption that economic growth was the prerequisite for achieving social cohesion and environmental protection (Scrieciu and Stringer 2008: 182). Urban regeneration, which emerged in this period as the fifth critical sphere of urban commonality, turned out to be an illustrative example of how the idea centred on improving quality of life in practice contributed to the rising social inequalities and injustice, through the related processes of commercialisation and gentrification.

Housing

The commodification of housing in CEE after 1989 resulted from the concurrent devolution of the system of state housing provision and marketisation of the housing sector. Following decentralisation, local authorities not only refrained from expanding municipal housing stock (Lux et al. 2003, Matoušek 2013), but also opted for privatisation of the existing units (Priemus and Mandi 2000, Tsenkova 2009). The privatisation strategy consisted of the restitution of property to its original owners or in a giveaway sale to sitting tenants—either free of charge or at a fraction of the market value. As a result, during the transition period, the proportion of public rentals decreased significantly in the whole region. Between 1990 and 2001, the share of public housing dropped by about one-fourth in the Czech Republic and Slovakia, by a half in Poland and Bulgaria, and by up to 80–90 per cent in Hungary and Romania (Hegedüs 2013: 15). Just before the global crisis of 2008, the shares of owner-occupied accommodation were above the EU-27 average in the new member states, with the sole exception of the Czech Republic (Czischke and Pittini 2007: 18).[12]

Both restitution to owners and sale to sitting tenants came with hidden costs incurred by the less privileged social groups. A toll connected to processes of restitution—typical of Warsaw because of the contemporary repercussions of the Bierut Decree—followed from insufficient legal regulations, which enabled dishonest private investors to acquire municipal tenements on the basis of flawed documentation (Polanska 2017). Dubbed 'savage reprivatisation', it often involved the new owners resorting to brutal and illegal methods to dispose of the building's residents (Springer 2015, Siemieniako 2017). A noticeable side-effect of sales to sitting tenants was the emergence of a vulnerable category of new owners, known as 'cash-poor and asset-rich' (Hegedüs and Teller 2006) or 'too poor to move, too poor to stay' (Lowe 2004), not being able to afford the rising utility and maintenance costs of their dwellings acquired almost free of charge.

All in all, large-scale reprivatisation and the accompanying retreat from the construction of new municipal housing units in larger cities increased the existing social inequalities by further limiting access to affordable housing for the 'worse connected' (Bodnár 1996, Kährik 2000, Lux 2001). Particularly negative was its impact on housing opportunities for the young (Roberts 2003). Consequently, housing lost its egalitarian quality of communal infrastructure. Losing its characteristic of a social right or entitlement, it rather quickly turned into a scarce commodity, or an asset, and an essential, although sometimes problematic, wealth reservoir. Unlike in the socialist period, the shelter function of housing became outweighed by the investment (property) function (Mandič 2010).

With reference to the lack of efficient housing policy in the transitioning countries of South-East Europe, Sasha Tsenkova (2009: 77) ascribes the leading role in the emerging housing market to the developers, the landowners, the financial institutions, and the building industry. All of them ranked higher in the hierarchy than the local housing and planning authorities, followed only by the most vulnerable group of actors—the consumers. The weakness of the regulatory institutions against the mostly private housing industry, which left those seeking accommodation and facing the rising prices almost entirely on their own, was characteristic of the whole former Eastern Bloc. It was also one of the factors contributing to the fast development of mortgage loan financing systems, which definitively redirected the housing economy toward the private sector. Housing cooperatives and other common ventures that were not compliant with the newly prominent, highly individualistic discourse of the '(sacred) right to private property' stood no chance in the race; they have remained at the margins of the housing production to this day (Coudroy de Lille 2015).

However imperfect the state housing policy was during the socialist period, the effects of its abandonment after 1989 have been deeply felt in the sprawling suburbs. What unleashed their development was the hunger for more affordable accommodation, coupled with relaxed urban planning regulations (Andrusz et al. 1996, Pichler-Milanovic et al. 2007). However, the quickly developing greenfield housing estates were usually poorly equipped with basic infrastructure—the much-needed nurseries, schools, as well as paved roads, and even pedestrian pavements. Furthermore, an increasing spatial disorder crept in, and the daily commuting, heavily dependent on individual transport, generated excess traffic congestion.

As outlined in Chapter 1, suburbanisation in CEE may be regarded as a by-product of increasing social atomisation and residential segregation (Węcławowicz 1998, Jałowiecki and Łukowski 2007, Sýkora 2009, Sagan and Marcińczak 2011, Marcińczak et al. 2012, Tammaru et al. 2015, Jaczewska and Gregorczyk 2017). Those who had followed the dream of a suburban home—often idealising living outside of the city or having little financial choice—found themselves without resorting to any existing community which they could call upon and in which they could self-organise (Kajdanek 2011). The sprawling of Central European cities in a way signals the weakening of social ties and citizen participation within urban arenas, an extreme case being gated communities (Gądecki 2009, Gąsior-Niemiec et al. 2009, Hirt 2012, Smigiel 2014). According to Stanilov (2007: 273), the astonishing popularity of gated communities in the post-socialist city 'underscores the failure of both the preceding communist regimes and the governments which followed to build an equitable and just society', as the more affluent residents '[s]eeking rescue from the chaos of the post-socialist reality, [...] have decided to build their own insulated version of personal paradise, letting the rest of the city crumble to pieces'.

Meanwhile, after the first wave of suburban flight, a renewed interest in the inner city surfaced in larger urban centres of CEE. Around the 2000s, newcomers began to infiltrate old, decayed neighbourhoods, regardless of their ongoing degradation. The residential attractiveness of these areas was boosted partly because of the planned regeneration schemes and partly as a consequence of wider socio-demographic and lifestyle changes (Haase et al.

2011, Kährik et al. 2015). The latter increased the significance of traditional pull factors, such as central location and the qualities of pre-1945 architecture, as well as recognised the value of the maintained social mix. At first inconspicuous, the influx of newcomers to the inner city sparked a debate on the threat of gentrification and the limits of application of this Western concept in the post-socialist context (Gądecki 2012, Chelcea et al. 2015, Górczyńska 2015, Grabkowska 2015, Kubeš and Kovács 2020).

Public transport

The neoliberal retreat of the state also took its toll on urban public transportation systems. During the 1990s and 2000s, the well-developed communal transport infrastructure in CEE crumbled under the weight of transition-related financial hindrances, the transformation of lifestyle and policy priorities, and the unconstrained urban spatial development. Although the booming suburbanisation demanded more effective public mobility networks, the loss of control over the process left the municipalities unable to keep up with the urban sprawl and resulted in poor integration between transportation and spatial planning (Stanilov 2007: 279). Dispersion and off-centre location of new housing districts were one of the factors causing a modal shift from public transport to private cars in larger cities after socialism (Pucher 1999, Komornicki 2003).

With the increasing employment in services and destandardisation of commuting patterns, automobility soon reigned throughout CEE. A formerly unattainable object of desire, the car counted not only as means of transport but also as a symbol of prestige and social standing (Pucher 1999, Komornicki 2003). Consequently, the demand for individual car travel in urban areas accelerated, unlike in Western Europe where it was beginning to slow down. The growth rate of the number of passenger cars varied across the countries of the former Eastern Bloc. It was the most rapid in the least motorised Romania, and relatively slower in the Czech Republic, Slovakia, and Hungary, where there had initially been more vehicles (Komornicki 2003). Automobilisation processes accelerated after the consecutive enlargements of the EU, partly through large-scale investments in roads and highways, which aimed to adjust the new member states to the European transport and cohesion policies (European Commission 2007, Rosik et al. 2015). At the local level, restricted public budgets put additional pressures on unprofitable municipal transport services, and the newly available EU funding was mostly directed to road-building programmes.

Initially, environmental issues were of lesser concern to the local decision-makers, who welcomed the partial transfer of responsibility for urban transport to the private sector (Mulíček and Seidenglanz 2019) and did little to discourage urban dwellers from individual car use. Not only did the implementation of parking fees proceed with reluctance, but the new road infrastructure often marginalised pedestrians and cyclists in favour of drivers (Nikšič 2017). Consequently, by 2012—one year before the congestion charge was first introduced in Central London—the number of registered vehicles per 1000 inhabitants in the 16 largest Polish cities ranged from 354.7 in the least affluent Białystok to 580.0 in the capital city of Warsaw (Parysek 2016: 14). Air pollution hit alarmingly high levels not just because of the rapidly growing number of vehicles, but also due to their poor technical condition (Coşciug et al. 2017, Kołsut 2020).

Alongside privatisation and individualisation, the reorganisation of urban transport unequally redistributed the related costs and benefits across the post-socialist societies. Some urban dwellers benefitted from increased mobility, enjoying the flexibility and comfort of car travel, and fulfilling their long-established aspirations regarding the status and social prestige of car ownership. The trade-off consisted of external effects manifest in transport-related social exclusion, suburban sprawl, heavy expenditure on road infrastructure, costs of road accidents, as well as air and noise pollution (Kronenberg and Bergier 2010, Gitkiewicz 2019, Trammer 2019).

The accumulation of these adverse factors pushed the cities to reconfigure their policies along the lines of sustainability. A particular effort was made to re-establish the role of rail transport. However, despite the available funding, many of the early attempts struggled or failed due to the lack of experience and/or compromised planning standards (Kołoś and Taczanowski 2018). In Gdańsk, selected as a co-host city of the 2012 UEFA European Football Championship, a new light railway line was planned to connect the football stadium with the local airport. After the event, the line was intended to serve as a commuting option for residents of the sprawling peripheral neighbourhoods, and thus relieve the increasing traffic congestion along one of the city's most overburdened communication axes. Due to a three-year delay in the construction works, the line was opened only in time to serve the latter function, originally intended as a 'follow-up'. However, its limited accessibility, exacerbated by insufficient integration with other public transport networks, called the effectiveness of the undertaking into question (Połom et al. 2018).

Another obstacle in the path to rehabilitation of communal transport infrastructures was related to the changing social attitudes towards collective mobility. Once preeminent, it became increasingly considered a necessary evil—overcrowded, unreliable, and too expensive in relation to the quality offered (Pucher and Buehler 2005). Moreover, its collectiveness was now perceived as a flaw. The lack of necessary reforms which could have improved this image added to the vicious circle, as the resulting passenger outflow reduced budget revenues and led to further decreases in passenger numbers. In post-socialist Sofia, even the popularity of the metro—the most valued means of collective transport, appreciated for its predictability and high standard—could not match the popularity of the car (Plyushteva 2019).

Green infrastructure

It is quite ironic that although environmental activists had had their fair share in the demise of the communist regimes in CEE, the green agenda was lost soon afterwards (Podoba 1998). Ecological concerns gave way to other priorities. Moreover, the social significance of urban green infrastructure, sanctioned by the former regime, was much ignored following 1989, since the collective character of recreation, leisure, and social interactions ceased to be ideologically relevant. A video art project about the transformation of the park in Chorzów juxtaposes archival photographs of local inhabitants participating in the construction works and using the completed facilities, accompanied by a narrative on the fullness of life, with shots visualising the emptiness and stillness of the park grounds at the beginning of the 21st century (Witkowska and Witkowski 2006).

The processes of deindustrialisation to some extent improved the quality of the natural environment during the early phase of the transition; however, these effects were insufficient and not supported by adequate policymaking (Haase et al. 2014). Even though environmental protection and management were later recognised as key aspects of urban planning in countries of the former Eastern bloc, the lack of comprehensive regulations has persisted to this day. The repetitive use of the empty slogan of sustainability has been often found to accompany cosmetic changes introduced in place of sweeping reforms which would be potentially unpopular or against the interests of influential lobbies (Podoba 1998).[13] Not infrequently, policymakers exhibited wishful thinking, establishing high standards without backing them up with holistic executive frameworks. For instance, due to the absence of proper regulations at the municipal level, quantitative norms regarding the provision of public urban green space per inhabitant, imposed in Romania after the country's accession to the EU, were not met by 32 out of 41 major cities (Luca et al. 2015).

By the time the economic value of greenery was acknowledged under new urban planning systems, the quantity and quality of urban green areas had already diminished. As shown in various studies from across the region, the total amount of greenery in CEE cities declined significantly in the 1990s and early 2000s (Badiu et al. 2019, Haase et al. 2019). This tendency was the most evident in residential areas. After 1989, unlike in the socialist era, the construction of new housing estates failed to take green infrastructure into account. Having little interest in their provision, developers followed the early capitalist standards, showing disregard for greenery in favour of building density. In central and inner-city locations, the rising real estate pressures threatened the existing urban parks, green squares, and, especially, overgrown plots of undeveloped land, which were often war relics.

Neither did the large housing estates escape the perils of privatisation and commercialisation. Vast expanses of open space intended for common use underwent dilapidation and/or were subjected to the intensification of residential development driven by individual, profit-oriented interests (Matlovič et al. 2001, Sendi et al. 2009, Kristiánová 2016, Szafrańska 2016). *Za Żelazną Bramą*, a centrally located large housing estate in Warsaw, built at the turn of the 1970s according to design principles proposed by Le Corbusier, may serve as a case in point. Forty years after its completion, the estate's neglected vegetation was intersected with concrete paths, fenced provisional parking lots, garages, and shackles. New residential developments implanted in-between the original buildings further affected the green infrastructure. Once a significant element of residential attractiveness, offsetting the small size of the flats, greenery has lost its value (Szynkarczuk 2015: 62).

An issue graver than the neglect was the threat of enclosure, engendered by the increasing fragmentation and privatisation of urban space. This particular factor contributed also to the rising pressure against the allotments, especially those situated in the most sought-after urban locations. Presented as 'relics of the socialist past', they would be seen as 'wastelands', or ineffectively used areas standing in the way of potential investment (Haase et al. 2019: 114). Likewise criticised were their supposed low aesthetic quality and semi-private character (Gibas and Boumová 2020). Despite the gardeners' efforts to self-organise, the number of allotments steadily decreased. In Prague, between 2004 and 2014, more than one-third of allotments had been lost due to abandonment, decline, or transformation into residential areas, while only less than a half of the remaining 261 maintained their original use, that is, gardening (Spilková and Vágner 2016). Even when the land cleared of the allotments was re-used for public, and not private, purposes, it would remain a desirable target for the latter. Such was the case of the Ronald Reagan Park, established between 2003 and 2006 in Gdańsk. Adjacent to the seashore, the area was transformed into open recreational grounds, featuring a system of drainage canals and ponds, walking and cycling paths, and several playgrounds. As the park gained popularity, much increasing the attractiveness of the nearby large housing estates from the socialist era, there was a flip side to its success. Namely, a plan emerged to develop fragments of the park into gated residential estates—two of which were eventually built amidst citizen protests (Sas-Bojarska 2015).

During the transition period, the perception of green infrastructure thus underwent two decisive shifts. At first, its value, assessed with regard to the purely recreational function, was depreciated and overlooked due to the prioritisation of other, more cost-effective uses of urban space. Next, it became acknowledged as one of the key factors for increasing residential attractiveness, sustainability, and quality of life (Badiu et al. 2019), and, as such, a profitable asset in land management. As a result, urban policies and developers' objectives began to change, slowly incorporating the green agenda into urban planning.

Public space

The civic spirit that led 'the people' to reappropriate urban public space expired shortly afterwards. To cite Stanilov (2007: 270), '[g]radually, the romanticism of the revolution and the political impetus it generated were pushed away by the relentless advance of the pragmatic forces of capitalism epitomized by two main processes—privatization and commercialization'. The former translated into the reduction of available public spaces, due to restitution processes and massive selling of communal land.[14] The latter was connected to the qualitative rearrangement of urban space in accordance with the principle of profit maximisation for private actors. Both types of transformations were often interlinked and resulted in further undermining of the publicness of urban space. Just like before 1989, the control over public space remained practically out of citizens' reach.

Some examples of privatisation of public space have been recalled already—they include the transfer of spaces between buildings in the socialist housing estates and fragments of urban parks into the hands of developers. In these instances, the new boundaries between the public and the private were set clearly. Some other changes were fuzzier, as in the case of Privately Owned Public Spaces (POPS), a phenomenon completely novel in post-socialist cities. Private ownership of the renowned Potsdamer Platz in Berlin is concealed under a deceptively welcoming spatial design, but as long as visitors abide by the unwritten rules. John Allen (2006: 445) describes how such an arrangement, aiming '*both* to encourage *and* to inhibit how we move around, use and act within' public space, accounts for indirect exertion of control over it through 'ambient power'.

The case of Potsdamer Platz also depicts how the production of public space succumbed to commercialisation. After several decades of oblivion during the Cold War period, the square reincarnated as a new type of marketplace, now mostly filled with tourists on their way to snap an obligatory selfie with the tent-like roof of the Sony Centre in the background. Under the new socio-economic order, palaces of culture were replaced with palaces of commerce. The shopping centre in particular became a trademark of the post-socialist city. Judit Bodnár (2001: 147) even goes as far as to state that the latter 'is built in the form of shopping malls'. She also maintains that the transition from limited consumer choice under state socialism to an abundance of consumer goods in post-socialist capitalism largely contributed to the uncritical reception and popularity of the new retail facilities after 1989, far outdoing their Western counterparts in performing the role of quasi-public spaces. A blatant example of this excess may be found in the recent trend of redeveloping railway stations in larger Polish cities into de facto shopping centres (Dragan 2017). Aware of its atomising effect, Gregory Andrusz (2006) observes that 'the materialistic values that the Mall embodies cannot serve as the glue that binds shoppers to one another as citizens' (2006: 85).

In an especially multifold manner, commercialisation processes affected public spaces in the inner city. Owing to the restored mechanism of the land rent, the housing function, so far reigning in centrally located neighbourhoods, had to give in to other uses, more profitable and compatible with the growing services sector. The remodelling of Budapest's Moscow Square exemplifies the shift in activities and décor towards 'commerce in its various manifestations' (Bodnár 2001: 104). They involved the emergence of undocumented immigrant workers awaiting potential employers and the proliferation of street vendors laying out flowers, handiwork, or cheap mass-produced goods for sale, but also the arrival of Burger King and McDonald's, along with a profusion of advertisements, signs, and information boards. Referring to Polish cities, Piotr Lorens (2016: 255) claims that they have been particularly susceptible to similar transformations due to the functional atrophy of their historic centres, resulting from the postponed development before WWII and aggravated during the state-socialist period.

The replacement of socialist murals, posters, and neon signs with commercial ads accounts for one of the numerous forms of post-socialist cleansing of public space. As it had happened in the post-war decades, symbol-laden street names were once again rebranded, either through de- or re-commemoration (Azaryahu 1997, Light 2004), while many other ideological elements of an urban landscape—Soviet monuments, red stars and flags, hammer-and-sickle emblems—were simply removed (Foote et al. 2000). Whichever of the cleansing strategies were put into practice after 1989—'removal, renaming, rededication or just reuse' (Czepczyński 2008: 115)—the result would be much the same. Once released from the ideological burden of communism, the post-socialist public space became instantly harnessed by consumer capitalism.[15] While the impetus for commodification was given by free-market mechanisms, the process was next incorporated into public strategies of place marketing and urban reinvention (Colomb 2012, Dellenbaugh-Losse 2020). Under the auspices of local authorities, new public spaces would be (re) designed and (re)developed with a specific type of user in mind. Targeted at audiences rather than participants, middle class rather than everyone, they favoured the have over the have-nots and prioritised individual profit-making over more communal objectives.

All in all, during the transition, urban policies at the local level in CEE were not aimed at restoring the publicness of public space. On the other hand, to paraphrase Podoba's (1998: 142) stance on environmental matters, the ambience of the privatised and commercialised public space effectively discouraged people from taking part in civic life. Both these tendencies were similarly pronounced in urban regeneration projects.

Urban regeneration

While housing, public transport, green infrastructure, and public space constituted four pillars of urban commonality in CEE under socialism, urban regeneration became the attribute of cities in and after transition. On the one hand, this was necessitated by the scale of inner-city degradation during the socialist era. Lack of investment in pre-war tenement buildings—due to financial constraints as much as ideological reasons—translated into physical dilapidation of the urban fabric and the deterioration of living conditions in old neighbourhoods. This led to the flight of better-off residents after 1989, and, consequently, to a concentration of social and economic problems. On the other hand, the forced deindustrialisation in the early 1990s and 2000s relieved vast brownfield areas formerly taken up by industrial and military facilities, often located close to the city centres (Frantál et al. 2013, Bosak et al. 2020).

Although not an entirely new idea, urban regeneration in a post-socialist city is a novelty in at least two respects. Firstly, while post-war reconstruction and later big-scale interventions, such as the demolition and consecutive rebuilding of inner-city Bucharest under Ceaușescu, fall under the broad definition of urban regeneration, they do not correspond to the narrow understanding of the term, that is, a comprehensive and integrated programme directed toward the improvement of economic, physical, social, and environmental conditions (Roberts 2000). In this regard, the holistic approach modelled after regeneration strategies applied in Western cities emerged as an innovative modus operandi. Secondly, the emancipation of local governments required consistent planning policies and the provision of sufficient funding. Lacking both adequate experience and financial resources, cities found it difficult to bring their visions to reality.

In the beginning of the transition period, urban regeneration was thus very much limited to skin-deep renovations in the inner city. The applied measures mostly concerned aesthetic makeovers or the improvement of technical conditions (Figure 2.3). For instance, in Banská Bystrica, a small Slovakian town of mediaeval origin, the transformation of the city centre involved mostly spatial changes. In 1994, the main square was pedestrianised and repaved, with elements of the pre-war look restored and accompanied by new street furniture. Although citizens expressed approval of the metamorphosis, which grew into 'a symbol of internationalization, westernization and Europeanism' (Bitušíková 1998: 619), criticism came from foreign tourists. In fact, it concerned the same features that had received praise from the locals, the argument being that the square lost its authenticity and now resembled other generic squares in Western Europe.

(a) (b)

Figure 2.3 City as a commodity: the main street in Łódź, Piotrkowska, after renovation (a) and advertisement in the window of one of the numerous banks located there (b).

In bigger cities, financial constraints propelled local governments to forgo 'unprofitable' community-oriented interventions and reach out to private investors (Masik 2012). The resulting large infrastructural projects carried out in public–private partnerships opened the door to the neoliberal agenda. Sylwia Kaczmarek and Szymon Marcińczak (2013: 105) recognise 'the strong belief in the effectiveness of liberal market forces and private property rights' as the most serious constraint for successful post-socialist urban regeneration. Even after the EU structural funds became available to the new member states, the policies of local authorities did not change substantially, at least not at first. In Poland, the long and winding road to socially focused revitalisation ran through culture-led development and the megaproject approach (Galuszka 2017). Ready-made formulas representative of both types were also applied in the rest of the region. They involved revisiting and reusing the potential of the cultural and industrial heritage, flagship urban interventions, as well as organising mega-events (Murzyn 2006, Temelová 2007, Slach and Boruta 2012, Hudec and Džupka 2016, Bosák et al. 2020). This entrepreneurial model, largely ignoring the actual needs of existing residents in terms of quality of life and the spatial coherence between the old and new development, resulted in regeneration by implantation rather than by integration (Kaczmarek 2001). A comprehensive comparative study of 56 regeneration projects located in 12 countries of the former Eastern Bloc and realised between 1989 and 2011 demonstrated the predominance of top-down approaches and low involvement of local communities in their implementation (Hlaváček et al. 2016).

The neglect of social cohesion in regeneration programmes observed throughout CEE also consisted of insufficiency or a complete lack of anti-gentrification measures. At the beginning of the 2000s gentrification—not yet a notable factor in the transformations of urban space—was limited to 'small islands in a wider sea of stagnation, decline as well as other forms of revitalisation' (Sýkora 2005: 104). With time, however, its consequences—especially with regard to displacement caused by new-build gentrification and state-led regeneration programmes—became increasingly felt (Kovács et al. 2013, Holm et al. 2015, Jakóbczyk-Gryszkiewicz 2015). Given the fact that citizens were excluded from the planning and implementation of the regeneration schemes, these social costs were unavoidable (Keresztély and Scott 2012, Sagan and Grabkowska 2012, Jarczewski and Kułaczkowska 2019).

City as a commons: return to the idea(l) of urban common good in the mid-2010s

Thirty years into post-socialism, urban stakeholders of the CEE seem to have gained more critical insight into the workings of the new socio-economic system. While citizens in Western Europe reached for the paradigm of the city as a commons after the disillusionment with the false promises of the neoliberal agenda, the discontent of Central and Eastern Europeans was aggravated by the previous failure of the state socialist urban project. On the one hand, there was a strong prejudice against urban commonality, and on the other—a certain nostalgia after some of its positive aspects and a lot of experience to draw from.

In retrospect, the city as a communal infrastructure appears today as an appealing approach,[16] only missing the crucial element of citizen participation in decision-making. The need for the latter in all aspects and phases of the planning process is enumerated by Zorica Nedović-Budić (2001: 47) among major planning and urban development issues in Eastern and Central Europe at the turn of the 21st century. This explains why throughout the region a comeback of the communal way of thinking about urban space was inspired by militant activists and informal groups of citizens determined to exercise their right to the city. As observed by Michaela Pixová (2020: 31), the much more pronounced, in comparison with the West, undemocratic nature of urban processes in Czech cities, affected the way they were addressed by both groups of urban actors.

Drawing from global and own past experiences, CEE citizens had thus re-discovered the common good at the local level and instilled collective action. For instance, in Poland urban activism and civic engagement gathered momentum in the early 2010s with the popularisation of the right to the city narrative through the Congress of Urban Movements (see Chapter 3) and in the 2014 campaign for the local government elections. However, as maintained by Katarzyna Kajdanek and Jacek Pluta (2016: 120), the legitimacy of public participation 'does not depend solely on the capability of social actors to employ practical and discursive knowledge in exerting political pressure on decision-makers'. Empowerment of the citizens, together with their ability to identify their own interests and needs, is essential for the institutional change to take hold. This disposition complies with the paradigm of city as a commons, which has been embraced in the CEE in spite of the conceptual disparities between commoning in the East and in the West (see Chapter 1). As it has been exercised in all five thematic threads discussed so far, it also pertains to an additional one. The bottom-up call for increasing quality of life in the city combined with minimisation of social inequalities and exclusion contributed to the emergence of spatial justice as the sixth area of articulation of the urban common good in the public sphere.

Housing

A former backbone (and unfulfilled promise) of public policies under socialism, left mainly to the market forces after 1989, housing in CEE became a central issue again in the first decade of the 21st century. Even though housing shortages had been largely resolved by then, this came at a cost of affordability. Not only the market prices of accommodation have remained relatively high, but the share of social rental housing has been insufficient in most of the countries in the region (Pittini et al. 2015). Following the transition, flats thus continued to be out of reach for many Central and Eastern Europeans—a source of contention which fed both the post-communist nostalgia and populist sentiments.

The first actors to respond to the injustices of the new system were tenants' movements. In Hungary, the Czech Republic, and Poland, associations aiming to safeguard and secure the rights of public renters in the face of the changing socio-economic situation started to form already in the late 1980s and early 1990s (Pickvance 1994, Pluciński 2014, Pixová and Sládek 2017). Although not massive in terms of membership, and with their message remaining on the margins of the public discourse, they consequently acted on behalf of the most vulnerable groups. In Poland, during the first two decades of transition, the image of tenants in the media was highly negative (Polanska 2017). Pictured either as demanding, unwilling to pay their rents and questioning the new economic order, or as 'losers' and victims of the transformation, they were pushed into the anti-heroes category. Alliance with radical anarchist organisations and squatters—in a campaign under the slogan 'Housing is a human right, not a commodity'—did not improve this unfavourable portrayal but helped to move their agenda forward (Jezierska and Polanska 2018).[17] In Hungary, Romania, and the Czech Republic, comparable processes contain an additional factor of ethnic discrimination towards the Roma population (Pixová 2020, Florea et al. 2018).

It is thus mainly owing to the grassroots urban movements, that the mainstream discourse has become disputed and countered with a narrative acknowledging the right to housing as a basic right alongside the rights to education or public transport. Inadequacies and failures of governmental housing programmes significantly contributed to rising expectations of the public reclaiming affordable accommodation as a social service. In Poland, the first attempts at such programmes, held from the mid-2000s to mid-2010s, were limited to subsidising mortgages which contributed to further inflation of the housing bubbles. This way, the state supported the developers more than the actual beneficiaries (Groeger 2016).

While residential preferences in CEE continue to be dominated by strong attachment to private property, alternative models of affordable housing have become increasingly appealing to urban dwellers. Evidence for development of options such as cohousing and cooperatives, although so far small-scale, may be found especially in larger cities (Czischke and van Bortel 2018, Twardoch 2019, Horňáková and Jíchová 2020). Joanna Erbel (2020) interprets this shift of attention as a symptom of housing ceasing to be perceived as an individual resource and becoming, again, a part of the common.

Public transport

At around the same time that the slight turn towards collectivity occurred in the domain of housing, signs of revival and rehabilitation of public transport could be observed in CEE cities. To some extent, they were brought about by the EU policymaking towards achieving sustainable urban mobility in member states (European Commission 2009, 2011, 2013). In the post-socialist context, the most challenging issues in this area have concerned limiting urban sprawl and reliance on individual car transport (Hołuj 2017), but also disrupting the already ossified aversion of urban dwellers to public transport (Radzimski and Gadziński 2019). Another problem is connected to uneven opportunities of urban centres correlated with their size. For instance, after the transitional hiatus, light-rail systems have developed quite well in the largest Polish agglomerations, as opposed to medium cities, which lack efficient substitutes to individual car transport (Kołoś and Taczanowski 2016).[18]

In general, local authorities throughout the region have been reluctant to implementing decisive policy measures reversing the urban transportation trends—not unlike in the rest of the EU (May 2015). Meanwhile, the sustainability agenda has been set bottom-up by urban activists and local civic organisations through public lobbying and campaigns. By way of an example, before the local government elections of 2018, one of the Polish think-tanks issued a report with recommendations addressed to candidates for mayors, calling for the implementation of best practices well-known from the West—limitation of individual car traffic in city centres, prioritisation of public transport, as well as increased investment in cycling and walking infrastructures (Górski 2017).

The gradually rising citizen awareness of the hidden environmental and social costs generated by transport-related pollution has been coupled with the transformation of urban lifestyles, preferences, and motivations. New patterns of urban transportation behaviour show that users' decisions to choose public modes of transport over cars tend to be based on such factors as the possibility of social interaction or environmental concerns (Jaśkiewicz and Besta 2014). Both conditions also account for the rising recognition of bicycle, used not only for recreation but also for moving around the city. Evidence from Sofia demonstrates a similar appreciation of urban commonality:

> rather than a straightforward withdrawal from the public domain, the growing popularity of cycling reveals a dynamic geography of inhabiting, sharing, contesting and enjoying the city's public spaces.
>
> (Barnfield and Plyushteva 2016: 1823)

Recent years have seen an expansion of transport sharing options in cities of CEE, from carpooling (Ivan 2010) through low-emission public transport (Taczanowski et al. 2018) to car-, bike-, and scooter-sharing (Kwiatkowski 2018, Pieriegud 2020). Although both their sustainability and potential for rapid development have been questioned, largely due to their dependence on coal-based electricity and low availability of charging infrastructures (Makowski 2017, Pieriegud 2020),[19] their proliferation substantially expanded available mobility options alternative to the car.

What accompanies these innovations are possible links to participatory practices and commoning in the post-socialist city. Within the former, citizens both provide local knowledge in the planning of urban transport infrastructures such as bike lanes (Niţă et al. 2018) and propose their own visions, as in the case of the first Polish woonerf in Łódź implemented via participatory budgeting (Kopeć and Wojtowicz 2021). An example of the latter is a bottom-up system of urban bike-sharing and repairing in Bratislava (Dellenbaugh et al. 2020).

Green infrastructure

Regardless of the eventual recognition of green spaces as relevant elements of urban infrastructure in CEE cities during the transition period, the tensions between development and preservation either remained high or continued to increase. On the one hand, the amount of urban land available for residential construction kept diminishing. Reurbanisation tendencies may have released the pressure on greenfields at the urban fringe but increased it within the inner city. On the other hand, issues over urban greenery and climate change reached and engaged wider audiences and impacted decision-making as a result. Greening regulations and programmes implemented in recent years by local authorities tend to apply participatory approaches (Belčáková et al. 2019, Lorencová et al. 2021).

While the initial bottom-up impulse for change came from environmental initiatives and movements, regular citizens became more active in exercising their right to a green city. Furthermore, there has been a qualitative change in the narrative. Moving away from a position of protest typical of the transition period, both groups of stakeholders increasingly turned to the adoption of the common good as a constructive argument in the debate. The term was invoked by the social side of the conflict over the gated residential investment planned on the edge of Reagan Park in Gdańsk during a presentation of an alternative vision of the area's development—aimed at protecting the existing landscape and biodiversity and maintaining it as an open urban garden (Sas-Bojarska 2015: 191).

Corresponding examples of re-communalisation of post-socialist urban green infrastructures concern the 'opening' of allotment gardens. In Prague, the initiative to make one of them accessible to the public was a reaction to the local authorities' plans to convert it into a public park which, apart from being against the interests of gardeners, posed similar threats as encountered in Gdańsk (Spilková and Vágner 2016). Analogous strategies have been adopted in other cities across the region, not just to pre-empt possible endeavours of the local governments but also due to embracement of their innovative and integrative character (Tóth et al. 2018, Mokras-Grabowska 2020). This argument is further supported by a boom of community gardens experienced in CEE since the 2010s (Bitušíková 2016, Maćkiewicz et al. 2018, Bende and Nagy 2020, Škamlová et al. 2020). A comparative study looking into the motivations of community gardeners in Bratislava, Budapest, Prague, Warsaw, and Zagreb found them to be connected to solidarity, collaborative work and the feeling of integration (Trendov 2018).

The described processes and interventions argue in favour of civic participation in the planning and management of urban greenery. While grassroots engagement often appears to be effective, there is a need for a more systemic approach. Moreover, as asserted by Alexandru Gavriş and Claudia Popescu (2021: 36), the issue of the lack of mutual trust which erodes the interactions between urban stakeholders and stands in the way must be overcome by means of 'a connected, collective empowerment'.

Public space

The evident 'fall of public space' in the post-socialist city, following its commodification and crisis of publicness, elicited responses from various milieus—urban movements and local communities, but also urban planners and practitioners. These different perspectives have operated in unison when criticising low aesthetic quality of urban spaces, their commercialisation and related exclusiveness, or incompatibility with the actual needs of citizens and other urban users. A narrative of the need for recovery of public space has also resonated in the media, both traditional and social (Grabkowska 2018). What made these claims go beyond the typical finger-pointing was the underlying readiness for taking action, accounting for 'a part of the broader process of regeneration of civic responsibility' (Czepczyński 2018: 70). It is therefore mostly in relation to public space that the notion of *com-munis*, forlorn under socialism and missing from the neoliberal agenda, has returned to the fore of public debate within the post-socialist counterpublic sphere.

The impact of new technologies has been apparent in this changeover as it facilitated informal social communication and collective action towards the urban common good (Grabkowska et al. 2013, Macek et al. 2015). In Łódź, the social media and mailing lists were the key mobilisation tools adopted by an anonymous citizen initiative going by the name A Group of Certain People (*Grupa Pewnych Osób*). Arranging and coordinating social interventions in public space by means of carnivalesque formulas of guerrilla gardening or fly-posting removal actions, they aimed not only at drawing the attention of the local authorities to some pressing deficiencies of the local public space but also at raising awareness among urban dwellers. Web-based but far from being limited to clicktivism, civic engagement within the Group over time moved from contestation and watchdogging towards co-governance—Certain People initiated public debate, and took active part in the local decision-making and community-led projects and events (Martela 2011). At the same time, despite the growing popularity and scale of the undertaken activities, the initiative refrained from institutionalisation to underline its egalitarian and non-hierarchic character (Przyłęcki 2015).

Bottom-up transformations of public space in the post-socialist city also concern old and new spatial practices. In the early 2010s in Poland, the most common forms of urban dweller's activity in the public space did not include civic but ludic and religious events, such as urban fests or Corpus Christi processions (Bierwiaczonek 2016: 151). Yet, repetitive protesting against

Figure 2.4 City as a commons: protest against controversial judiciary reform outside the local court in Kraków (20 July 2017).

violations of the constitutional order and the abortion ban, taking place since 2015 and 2016, respectively, have mobilised hundreds of thousands of citizens expressing discontent in the streets (Karolewski 2016, Korolczuk 2016) (Figure 2.4). Participative and discursive place-making thus went hand in hand with reclaiming civic landscapes (Czepczyński 2018). Recognition of post-socialist public space as a commons is also evident reclamation of urban streets and other open spaces by the formerly 'invisible', marginalised groups of citizens—women, LGBT communities, ethnic minorities, and people with disabilities (McGarry 2016, Lelea and Voiculescu 2017, Majbroda 2018, Kubicki et al. 2019).

It is important to underline that the performance of NGOs and informal spectacular mass mobilisations have not been the only modes to restore the lost commonness of public space in the region. Not denying the relevance of either, Kerstin Jacobsson and Elżbieta Korolczuk (2020) urge the acknowledgement of the unseen and undervalued role of 'uneventful activism'. While it is often 'low-key, small-scale, and initiated by individuals or small, informal groups, and little discussed in the mass media and public discourse' (130), it translates into the actual change of perception and use of the public space, as well as its partial decommodification.

Urban regeneration

Although the 'long-term problem' (Hlaváček et al. 2016: 38) of the low level of participation of local communities in the planning and implementation of regeneration projects in CEE remains unsolved, as of late there have been some signs of progress in this domain. In Poland, the Act on Regeneration, adopted in 2015 and filling the lacuna in the legal framework, prioritised the position of local communities as key stakeholders and beneficiaries.[20] Furthermore, the local authorities are obliged to carry out regeneration 'in an open and transparent manner, ensuring active participation of stakeholders at every stage', as well as 'in a way that prevents exclusion of local residents from benefitting from the positive effects of regeneration, with particular regard to the use of communal housing resources' (Art. 3.2). Two separate chapters are devoted to methods of public participation and establishment of the institution of the Committee for Regeneration.[21] While the effectiveness of these regulations remains to be seen, a recent comparison of urban regeneration processes in Poland and Bulgaria pointed to an advantage of the Polish cities in terms of citizen involvement (Ciesiółka and Burov 2021).

Independently from the legislative changes at the central level, in some Polish cities, engagement of local communities was embedded in regeneration strategies via communicative planning and the so-called 'participatory budgeting 2.0'. The former model underlies the successful transformation of a densely built inner-city area of Stare Polesie in Łódź. Aimed at greening the neighbourhood and improving pedestrian safety, it had commenced with a series of meetings with local residents dedicated to a thorough analysis of their needs and preferences and continued with their active involvement in the regeneration process (Zdyb 2017). The latter is best exemplified by a deliberative approach endorsed in Dąbrowa Górnicza, departing from the conventional rules of participatory budgeting adopted in Polish cities (Majorek 2018). Instead of the individualised 'submit & vote' formula, here the ideas are first debated and jointly elaborated during open public workshops, yielding informed and tailor-made proposals.

The described practices abandon the idea of regeneration reduced to revalorisation of space and/or economic revival, compliant with the city as a commodity paradigm. In my previous book I observed this shift taking place in Gdańsk in connection to bottom-up regeneration linked to reurbanisation—as soon as the newcomers to dilapidated inner-city tenements have secured private interests through renovation of their own flats, they tended to inspire incumbent upgrading of the entire buildings and engage in common initiatives for the sake of the common good at the level of the whole neighbourhood (Grabkowska 2012). Although accompanied by marginal gentrification, these community-led actions overall contributed to considerable social integration and increased sense of responsibility for common space, fulfilling the ideal of city as a commons.

Regardless of the harbingers of citizen engagement in urban regeneration in CEE, its inclusiveness has been repeatedly questioned on the grounds of domination of the middle class. For example, throughout the consecutive phases of regeneration of inner-city Magdolna Quarter in Budapest involvement of the local residents increased enough for it to be hailed as a 'best practice community participation project', but also criticised for exclusion of the more disadvantaged groups (Keresztély and Scott 2012, Nzimande and Fabula 2020: 394). This and other instances of the new bourgeoisie taking over the post-socialist counterpublic substantiates the need of investigating another area of urban commonality, that is, spatial justice.

Spatial justice

Understood as a combination of social and environmental justice in the context of and within urban space, the idea of spatial justice is inextricably linked to urban commonality. Without taking it into consideration, both the right to the city rationale and the paradigm of the city as a commons lose their raison d'être. However, as various studies show, it remains one of the greatest and insufficiently addressed challenges in Central and Eastern European cities after socialism. Even when attempts at resolving unequal access to urban common goods, and/or lack of agency at the local level, are made, they are unsuccessful due to systemic or cultural restrictions. Often no such attempts are made at all.

Given its cross-sectional characteristic, the distributive and procedural aspects of spatial justice appear in all five dimensions discussed so far within Chapter 2. Within housing issues, this notion has resonated in the discussion over privatisation of communal resources and the contested 'privileged' position of sitting tenants (Polanska 2017). However, it also comes up in the debates on the right to affordable accommodation for all. Injustices of public transport have mostly concerned the need for reclaiming the urban streets by the marginalised pedestrians and cyclists (Udvarhelyi 2009), as well as for the provision of efficient and reasonably priced public transport options. Controversies arose around the question of whether municipalities should invest in the transport infrastructures resultant from uncontrolled urban sprawl fed by developers. Environmental justice comes to the fore when we turn to the problems of urban green infrastructure. Although Kronenberg et al. (2020: 3) find it to be 'barely addressed or known outside narrow academic circles and environmental NGOs', in post-socialist countries other than former East Germany, it has been decisively on the rise, as proven by the review of scholarly works presented in the same paper. One of them, by Harper et al. (2009), is about the Roma communities who were declined the right to environmental safety and resources in Hungary and Slovakia, as a result of their 'misrecogni[tion] as indifferent to environmental issues and exclu[sion] from participating in public decision-making' (254). Critical debates on spatial justice relating to public space transformations concern such seemingly disparate issues as pressures of touristification on local residents' quality of life (Pixová and Sládek 2017, Kruczek 2018, Kowalczyk-Anioł 2019) and homelessness as 'alternative politics of living in current post-socialist cities' (Ferenčuhová and Vašát 2021) but in effect boil down to the question 'whose right to public space?'. Finally, the picture is completed by the aforementioned unfairness of urban regeneration.

In terms of the procedural dimension of spatial (in)justice in post-socialist cities, even the diffusion of participatory budgeting, promoted as a tool for community engagement, did not significantly help to improve the situation. Implemented in Bulgaria in the early 2010s, followed by the Czech Republic, Poland, Romania, and Slovakia, and, a bit later, in Hungary, in most urban locations it is project-based, relying on a competition between highly individualised and undebated proposals which goes against the original idea from Porto Alegre (Martela 2020, Klimovský et al. 2021). Such a model of participatory budgeting accounts for 'wide but shallow consultations' (Mączka et al. 2021: 486), but also, perhaps, more importantly, it becomes a 'participatory tool which does not guarantee participation' (Siciarek 2014). Accordingly, it may even reproduce the existing power relations and hierarchies between the privileged and the marginalised urban stakeholders. This lesson also dispels the misconception that the emergence of the city as a commons paradigm makes it enough to secure equal rights for all potential commoners.

Notes

1 For an in-depth analysis of adequacy of the term 'civil society' in the Central and Eastern European context before 1989, see Gagyi and Ivancheva (2019).

2 The anti-democratic leadership of the Putin era has not challenged the validity of this diagnosis (see Langenohl and Schmäing 2020, Turovets 2020).

3 Instead, he literally invited the proletariat to move into downtown Warsaw. According to his words, the construction of centrally located workers' housing would symbolically eradicate 'the capitalist tradition of pushing out the working class to the suburbs' (231). For a poetic interpretation of this political vision, see Adam Ważyk's (1950: 107) social realist verse 'Lud wejdzie do śródmieścia' ('The people will enter the city centre').

4 In Polish People's Republic, citizens were legally obliged to 'safeguard and strengthen social property' defined as 'the unshakable foundation of development of the State, source of wealth and might of the country' (Constitution of Polish People's Republic 1952: art. 77.1). Failing to comply with this regulation would be punished (art. 77.2).

5 The state socialist model in fact pertains to any imposition of collectivism effectively sterilising the public sphere. Illustrative here are twin examples of atomic cities from each side of the Iron Curtain—Richland in the US and Ozersk in the USSR—portrayed in Kate Brown's *Plutopia* (2013). They both emerged as top secret, isolated urban projects built to accommodate the employees and families of strategic plutonium plants during the Cold War. In both cases, the elements of urban common good—privileged but powerless inhabitants, advanced communal infrastructure, and forced egalitarian communality—remained in a static relationship, producing rather dystopian urban environments.

6 Although, to complete the picture, it must be added that anti-social behaviour, such as looting or mugging, also happened on a daily basis.

7 In Warsaw, critics of the Bierut Decree, which in 1945 communalised all land within the pre-war boundaries of the city, were sneeringly referred to as defenders of the sacred right to property (Piątek 2020: 174). In Romania, the goal of the 1950 housing nationalisation Decree 92 was 'to strengthen and develop the

socialist sector in Romania', 'to better administer the housing stock at risk of dilapidation because of the sabotage of the high bourgeoisie and exploiters who own a large number of buildings', and 'to deprive exploiters of an important means of exploitation' (Şerban 2014: 781).

8 The additional 'value'of this system was that it enabled a certain degree of control over citizens. A pinnacle of collectivity, its coerciveness nevertheless contradicted the idea of urban commonality.

9 On a similar note, Olga Drenda, author of a hauntological essay on Polish material culture at the turn of the 1990s, juxtaposes the 'grim, lunar landscape' outside the housing blocks with their internal 'hobbit burrow's cosiness'. The latter was attained by an accumulation of 'wood panelling, carpets and rugs, (…) tablecloths and serviettes, airy curtains and floor-length drapes, patterned upholstery, puffy leather-covered doors—anything that insulates, cushions, calms and soothes' (Drenda 2016: 60–61, see also the contributions of David Crowley and Katerina Gerasimova in Crowley and Reid 2002).

10 Two prominent exceptions to the rule include the Budapest Metro, which has operated since 1896 as the oldest in continental Europe, and the Berlin U-Bahn, completed in 1902. The former underwent considerable extension and modernisation from the 1950s onwards, while the development of the latter was hampered following the city's division by the Wall in 1961.

11 The common effort of the local community is embodied in the external walls decorated with pebbles collected and delivered by the parishioners from their trips all over the country.

12 In Germany, the level of private ownership was the lowest, with private rental being the highest in the EU.

13 Little effort made towards restructuring the energy sector (adherence to fossil fuels) is just one example of the persistence of environmentally costly paradigms.

14 On the microscale, it was expressed in the inclination of urban dwellers to isolate from each other in common space and appropriate its fragments for their own use. Typical of such behaviour is the phenomenon of 'windscreening' observed in the Polish seaside, where some beach users rush in the early morning to entrench themselves in the best spots.

15 Some components of the newly bygone urban past were spared, or made a swift return on the wave of Ostalgie (Berdahl 1999, Boyer 2006) or post-communist nostalgia (Todorova and Gille 2010). For instance, while the much-hated Berlin Wall was pulled down almost entirely, the only three remaining sections of the original structure acquired the role of tourist landmarks and became equipped with adequate infrastructure serving to attract visitors from all over the globe.

16 Manifestations of a global rise of interest in infrastructural issues within a contemporary city are discussed by Alan Latham and Peter Wood (2015: 300–301), who observe a renewed appreciation of urban infrastructures, awareness of their currently changing modes of use and management, and recognition of their social and political underpinnings going beyond technical conditions.

17 A decisive turn in the struggle for legitimisation happened only after the tragic death of a tenants' activist Jolanta Brzeska in March 2011, coinciding with a series of other reprivatisation scandals.

18 Reintroduction of a tramway line in Olsztyn in northern Poland has been so far a unique undertaking of this kind (Gadziński and Radzimski 2021).

19 Worldwide, dockless e-scooter systems are also criticised for posing safety hazards and generating disorder in urban space (Lipovsky 2021).

20 Its Article 2.2, listing all parties to be engaged in the regeneration process, features the following order: local residents, property owners and managers, residents from outside of the area, economic and social entities, local government units and public authorities.

21 Apart from the typical call for comments, the Act requires the consultation process to include at least two non-standard options available from an exhaustive list including, inter alia, debates, workshops, and study walks (Art. 3). Committee for Regeneration, 'a forum for cooperation and dialogue between the stakeholders' (Art. 7.1) is a consultative body for planning, implementation and evaluation of regeneration projects at the local level.

References

Allen J (2006) Ambient Power: Berlin's Potsdamer Platz and the seductive logic of public spaces. *Urban Studies*, 43(2), 441–455.

Andrusz G (2006) Wall and mall: A metaphor for metamorphosis. In: Tsenkova S, Nedović-Budić Z (eds), *The Urban Mosaic of Post-Socialist Europe: Space, Institutions and Policy*. Heidelberg: Physica-Verlag, 71–90.

Andrusz G, Harloe M, Szelenyi I (eds) (1996) *Cities after Socialism: Urban and Regional Change and Conflict in Post-socialist Societies*. Oxford: Blackwell.

Asen R (2017) Neoliberalism, the public sphere, and a public good. *Quarterly Journal of Speech*, 103(4), 329–349. https://doi.org/10.1080/00335630.2017.1360507

Azaryahu M (1997) German Reunification and the Politics of Street Names: The Case of East Berlin. *Political Geography*,16, 479–493.

Badiu DL, Onose DA, Niță MR, Lafortezza R (2019) From "Red" to Green? A Look into the Evolution of Green Spaces in a Post-socialist City. *Landscape and Urban Planning*, 187, 156–164. https://doi.org/10.1016/j.landurbplan.2018.07.015

Bajomi-Lázár P (ed.) (2017) *Media in Third-Wave Democracies. Southern and Central/Eastern Europe in a Comparative Perspective*. Paris: L'Harmattan Publishing House.

Barnfield A, Plyushteva A (2016) Cycling in the post-socialist city: On travelling by bicycle in Sofia, Bulgaria. *Urban Studies*, 53(9), 1822–1835.

Belčáková I, Świąder M, Bartyna-Zielińska M (2019) The green infrastructure in cities as a tool for climate change adaptation and mitigation: Slovakian and polish experiences. *Atmosphere*, 10, 552. https://doi.org/10.3390/atmos10090552

Bellows AC (2004) One hundred years of allotment gardens in Poland. *Food and Foodways*, 12(4), 247–276. https://doi.org/10.1080/07409710490893793

Bende C, Nagy G (2020) Community gardens in post-socialist Hungary: Differences and similarities. *Geographia Polonica*, 93(2), 211–228.

Berdahl D (1999) '(N)Ostalgie' for the Present: Memory, Longing, and East German Things. *Ethnos*, 64(2), 192–211.

Bierut B (1951) *Sześcioletni Plan Odbudowy Warszawy*. Warszawa: Książka i Wiedza.

Bierwiaczonek K (2016) *Społeczne znaczenie miejskich przestrzeni publicznych*. Katowice: Wydawnictwo Uniwersytetu Śląskiego.

Bitušíková A (1998) Transformations of a city centre in the light of ideologies: the case of Banská Bystrica, Slovakia. *International Journal of Urban and Regional Research*, 22, 614–622.

Bitušíková A (2015) Urban Activism in Central and Eastern Europe: A Theoretical Framework. *Slovenský Národopis/Slovak Ethnology*, 63(4), 326–338.

Bitušíková A (2016) Community gardening as a means to changing urban inhabitants and their space. *Critical Housing Analysis*, 3(2), 33.

Bodnár J (1996) 'He That Hath to Him Shall Be Given': Housing Privatization in Budapest After State Socialism. *International Journal of Urban and Regional Research*, 20(4), 616–636.

Bodnár J (2001) *Fin de Millénaire Budapest: Metamorphoses of urban life*. Minneapolis: University of Minnesota Press.

Bodnár J, Böröcz J (1998) Housing Advantages for the Better Connected? Institutional Segmentation, Settlement Type and Social Network Effects in Hungary's Late State-Socialist Housing Inequalities. *Social Force*, 76(4), 1275–1304.

Bosák V, Slach O, Nováček A, Krtička L (2019) Temporary Use and Brownfield Regeneration in Post-socialist Context: From Bottom-up Governance to Artists Exploitation. *European Planning Studies*, 28(3), 604–626. https://doi.org/10.1080/09654313.2019.1642853

Boyer D (2006) Ostalgie and the Politics of the Future in Eastern Germany. *Public Culture*, 18(2), 361–381.

Brzostek B (2007) *Za progiem: Codzienność w przestrzeni publicznej Warszawy lat 1955–1970*. Warszawa: Wydawnictwo TRIO.

Carmin J, Hicks B (2002) International triggering events, transnational networks, and the development of Czech and polish environmental movements. *Mobilization: An International Quarterly*, 7(3), 305–324. https://doi.org/10.17813/maiq.7.3.nv3226643761t786

Chelcea L, Druţă O (2016) Zombie Socialism and the Rise of Neoliberalism in Post-socialist Central and Eastern Europe. *Eurasian Geography and Economics*, 57(4–5), 521–544.

Chelcea L, Popescu R, Cristea D (2015) Who Are the Gentrifiers and How Do They Change Central City Neighbourhoods? Privatization, Commodification, and Gentrification in Bucharest. *Geografie*, 120(2), 113–133.

Cichońska I, Popera K, Snopek K (2019) *Day-VII Architecture: A Catalogue of Polish Churches Post 1945*. Berlin: DOM Publishers.

Ciesiółka P, Burov A (2021) Paths of the urban regeneration process in Central and Eastern Europe after EU enlargement – Poland and Bulgaria as comparative case studies. *Spatium*, 46, 1–10. https://doi.org/10.2298/SPAT2146001C

Colomb C (2012) *Staging the New Berlin: Place Marketing and the Politics of Urban Reinvention Post-1989*. Abingdon: Routledge.

Cordell K (1990) Political Change in the GDR: The Role of the Evangelical Church. *International Relations*, 10(2), 161–166. https://doi.org/10.1177/004711789001000205

Coşciug A, Ciobanu SM, Benedek J (2017) The Safety of Transnational Imported Second-Hand Cars: A Case Study on Vehicle-to-Vehicle Crashes in Romania. *Sustainability*, 9(12). https://doi.org/10.3390/su9122380

Coudroy de Lille L (2015) Housing cooperatives in Poland. The origins of a deadlock. *Urban Research and Practice*, 8(1), 17–31. https://doi.org/10.1080/17535069.2015.1011424

Crowley D (2002) Warsaw Interiors: The Public Life of Private Spaces, 1949–65. In: Crowley D, Reid SE (eds) *Socialist Spaces: Sites of Everyday Life in the Eastern Bloc*. Oxford: Berg, 181–206.

Crowley D, Reid SE (2002) Socialist Spaces: Sites of Everyday Life in the Eastern Bloc. In: Crowley D, Reid SE (eds), *Socialist Spaces: Sites of Everyday Life in the Eastern Bloc*. Oxford: Berg, 1–22.

Cymbrowski B (2017) The Proletarian Public Space and Its Transformation. The Case of Socialist and Post-socialist Cities. *Sociologica*, 2, 1–27. https://doi.org/10.2383/88199

Cymer A (2019) *Architektura w Polsce 1945–1989*. Warszawa: Fundacja Centrum Architektury.

Czapiński J (2015) Stan społeczeństwa obywatelskiego. In: Czapiński J, Panek T (eds), Social Diagnosis 2015: Objective and subjective quality of life in Poland. *Contemporary Economics*, 9(4), 332–372.

Czepczyński M (2008) *Cultural Landscapes of Post-socialist Cities: Representation of Powers and Needs*. Aldershot: Ashgate.

Czepczyński M (2018) Civic landscapes of post-socialist cities: Urban movements and the recovery of public spaces. In: Hristova S, Czepczyński M (eds) *Public space: between reimagination and occupation*. London: Routledge, 65–74.

Czischke D, Pittini A (2007) *Housing Europe 2007: Review of Social, Co-operative and Public Housing in the 27 EU Member States*. Brussels: CECODHAS European Social Housing Observatory.

Czischke D, van Bortel G (2018) An exploration of concepts and polices on 'affordable housing' in England, Italy, Poland and The Netherlands. *Journal of Housing and the Built Environment*, 1–21. https://doi.org/10.1007/s10901-018-9598-1

De Magistris A (2012) Architecture and urban planning. In: Ponce S, Service R (eds), *A Dictionary of 20th-Century Communism*. Princeton: Princeton University Press, 31–34.

Dellenbaugh-Losse M (2020) *Inventing Berlin: Architecture, Politics and Cultural Memory in the New/Old German Capital Post-1989*. Berlin: Springer.

Dellenbaugh-Losse M, Zimmermann N-E, de Vries N (2020) The Urban Commons Cookbook: Strategies and Insights for Creating and Maintaining Urban Commons. Retrieved from: http://urbancommonscookbook.com

Domaradzka A (2021) *Klucze do miast: Ruch miejski jako nowy aktor w polu polityki miejskiej*. Warszawa: Wydawnictwo Scholar.

Dragan W (2017) Development of the urban space surrounding selected railway stations in Poland. *Environmental and Socio-economic Studies*, 5(4), 57–65. https://doi.org/10.1515/environ-2017-0020

Drenda O (2016) *Duchologia polska: Rzeczy i ludzie w latach transformacji*. Kraków: Karakter.

Drozda Ł (2017) Pułapka gentryfikacji? Związki "uszlachetniania" przestrzeni z programami rewitalizacji polskich miast. *Studia Regionalne i Lokalne*, 4(70), 5–22. https://doi.org/10.7366/1509499547001

Dunn EC (2004) *Privatizing Poland: Baby Food, Big Business, and the Remaking of Labor*. Ithaca: Cornell University Press.

Engels F (1887) *The Housing Question*. Reprinted by the Co-operative Publishing Society of Foreign Workers. Retrieved from: https://www.marxists.org/archive/marx/works/1872/housing-question/index.htm

Enyedi G (1996) Urbanization under socialism. In: G Andrusz, M Harloe, I Szelenyi (eds), *Cities after Socialism: Urban and Regional Change and Conflict in Post-socialist Societies*. Oxford: Blackwell, 100–118.

Erbel J (2020) *Poza własnością: W stronę udanej polityki mieszkaniowej*. Kraków: Wysoki Zamek.

European Commission (2007) *Growing Regions, Growing Europe: Fourth Report on Economic and Social Cohesion*. Luxembourg: Office for Official Publications of the European Communities.

European Commission (2009) *Action Plan on Urban Mobility*. COM(2009). Retrieved from: https://transport.ec.europa.eu/transport-themes/clean-transport-urban-transport/urban-mobility/action-plan-urban-mobility_en490 (last accessed: 9 April 2022).

European Commission (2011) *White Paper Roadmap to a Single European Transport Area – Towards a competitive and resource efficient transport system*. COM (2011) 144. Retrieved from: https://eur-lex.europa.eu/LexUriServ/LexUriServ.do?uri=COM:2011:0144:FIN:en:PDF (last accessed: 9 April 2022).

European Commission (2013) *Urban Mobility Package. Communication "Together towards competitive and resource-efficient urban mobility*. COM(2013) 913. Retrieved from: https://eur-lex.europa.eu/legal-content/EN/TXT/?uri=celex%3A52013DC0913 (last accessed: 9 April 2022).

Ferenčuhová S, Vašát P (2021) Ethnographies of urban change: introducing homelessness and the post-socialist city. *Urban Geography*, 42(9), 1217–1229. https://doi.org/10.1080/02723638.2021.1930696

Fidelis M, Gigova I (2017) Communism and its Legacy. In: Livezeanu I, von Klimó Á (eds), *The Routledge History of East Central Europe since 1700*. London: Routledge, 365–414.

Florea I, Gagyi A, Jacobsson K (2018) A field of contention: Evidence from housing struggles in Bucharest and Budapest. *Voluntas*, 29, 712–724. https://doi.org/10.1007/s11266-018-9954-5

Foote K, Tóth A, Árvay A (2000) Hungary after 1989: Inscribing a New Past on Place. *Geographical Review*, 90(3), 301–334. https://doi.org/10.2307/3250856

Frantál B, Kunc J, Nováková E, Klusáček P, Martinát S, Osman R (2013) Location matters! Exploring brownfields regeneration in a spatial context (A Case Study of the South Moravian Region, Czech Republic). *Moravian Geographical Reports*, 21(2), 5–19.

Gądecki J (2009) *Za murami: osiedla grodzone w Polsce – analiza dyskursu*. Wrocław: Wydawnictwo Uniwersytetu Wrocławskiego.

Gądecki J (2012) *I ♥ NH: Gentryfikacja starej części Nowej Huty?* Warszawa: Wydawnictwo IFiS PAN.

Gadziński J, Radzimski A (2021) Impacts of light rail in a mid-sized city: Evidence from Olsztyn, Poland. *Journal of Transport and Land Use*, 14(1), 821–84.

Gagyi A, Ivancheva M (2019) The reinvention of 'civil society': Transnational conceptions of development in East-Central Europe. In: McCrea N, Finnegan F (eds), *Funding, Power and Community Development*. Bristol: Policy Press, 55–69.

Galuszka J (2017) Examining patterns of policy change in a post-socialist city: the evolution of inner-city regeneration approaches in Łódź, Poland, after 1989. *Town Planning Review*, 88(6), 639–66.

Gąsior-Niemiec A, Glasze G, Pütz R (2009) A Glimpse over the Rising Walls: The Reflection of Post-Communist Transformation in the Polish Discourse of Gated Communities. *East European Politics and Societies*, 23, 244–265. https://doi.org/10.1177/0888325408328749

Gavriş A, Popescu C (2021) The refeudalisation of parks: subduing urban parks in Bucharest. *Local Environment*, 26(1), 131–145. https://doi.org/10.1080/13549839.2020.1867838

Gentile M, Sjöberg Ö (2006) Intra-Urban Landscapes of Priority: The Soviet Legacy. *Europe-Asia Studies*, 58(5), 701–729.

Gibas P, Boumová I (2020) The Urbanization of Nature in a (Post)Socialist Metropolis: An Urban Political Ecology of Allotment Gardening. *International Journal of Urban and Regional Research*, 44(1), 18–37.

Giordano C, Kostova D (2002) The social production of mistrust. In: Hann CM (ed.), *Postsocialism: ideals, ideologies, and practices in Eurasia*. London: Routledge.

Gitkiewicz O (2019) *Nie zdążę*. Warszawa: Dowody na Istnienie.

Goldzamt E (1971) *Urbanistyka krajów socjalistycznych: Problemy społeczne.* Warszawa: Arkady.

Gömöri G (1973) The cultural intelligentsia: The writers. In: Lane D, Kolankiewicz G (eds), *Social Groups in Polish Society*. London: Macmillan Education, 152–179.

Górczyńska M (2015) Gentryfikacja w polskim kontekście: krytyczny przegląd koncepcji wyjaśniających. *Przegląd Geograficzny*, 87(4), 589–611.

Górski R (2017) Wstęp. In: Wesołowski J, Makowski W (eds), *Odzyskajmy centra miast*. Łódź: Instytut Spraw Obywatelskich, 7–9.

Grabkowska M (2012) *Regeneration of the Post-socialist Inner City: Social Change and Bottom-up Transformations in Gdańsk*. Gdańsk: Pracownia.

Grabkowska M (2015) Between gentrification and reurbanisation: The participatory dimension of bottom-up regeneration in Gdańsk, Poland. *Geografie*, 120(2), 210–225.

Grabkowska M (2018) Urban Space as a Commons in Print Media Discourse in Poland after 1989. *Cities*, 72, 122–129.

Grabkowska M, Pancewicz Ł, Sagan I (2013) The Impact of Web-Based Media on Evolution of Participatory Urban Planning and E-Democracy in Poland. *International Journal of E-Planning Research*, 2(3), 1–16.

Groeger L (2016) Programy wspierania budownictwa mieszkaniowego w Polsce i ich wpływ na rynek nieruchomości mieszkaniowych. *Space–Society–Economy*, 18, 131–146.

Haase A, Steinführer A, Kabisch S, Grossmann K, Hall R (eds) (2011) *Residential Change and Demographic Challenge: The Inner City of East Central Europe in the 21st Century*. Farnham: Ashgate.

Haase D, Dushkova D, Haase A, Kronenberg J (2019) Green infrastructure in post-socialist cities: Evidence and experiences from Eastern Germany, Poland and Russia. In: Tuvikene T, Sgibnev W, Neugebauer CS (eds), *Post-Socialist Urban Infrastructures*. London: Routledge, 105–124.

Haase D, Haase A, Rink D (2014) Conceptualising the nexus between urban shrinkage and ecosystem services. *Landscape and Urban Planning*, 132, 159–169.

Harper K, Steger T, Filčák R (2009) Environmental Justice and Roma Communities in Central and Eastern Europe. *Environmental Policy and Governance*, 19, 251–268. https://doi.org/10.1002/eet.511

Hatherley O (2015) *Landscapes of Communism – A History Through Buildings*. New York: The New Press.

Havel V (1990) Letter to Dr Gustáv Husák, General Secretary of the Czechoslovak Communist Party. In: Vladislav J (ed), *Václav Havel Living in Truth. Twenty-two Essays Published on the Occasion of the Award of the Erasmus Prize to Václav Havel*. London: Faber and Faber, 3–35.

Hegedüs J (2013) The Transformation of the Social Housing Sector in Eastern Europe: A Conceptual Framework. In: Lux M, Hegedüs J, Teller N (eds), *Social Housing in Transition Countries*. New York: Routledge, 3–30.

Hegedüs J, Teller N (2006) Managing risks in the new housing regimes of the transition countries: The case of Hungary. In: Doling J, Elsinga M (eds), *Home Ownership: Getting In, Getting From*, Getting Out. Part II. Amsterdam: IOS Press, 175–200.

Heller M (1988) *Cogs in the Soviet Wheel: The Formation of Soviet Man*. New York: Alfred A. Knopf.

Hirt S (2007) The Compact versus the Dispersed City: History of Planning Ideas on Sofia's Urban Form. *Journal of Planning History*, 6(2), 138–165. https://doi.org/10.1177/1538513206301327

Hirt SA (2012) *Iron Curtains: Gates, Suburbs and Privatization of Space in the Post-socialist City*. Oxford: Wiley-Blackwell.

Hlaváček P, Raška P, Balej M (2016) Regeneration projects in Central and Eastern European post-communist cities: Current trends and community needs. *Habitat International*, 56, 31 41.

Holm A, Marcińczak S, Ogrodowczyk A (2015) New-build gentrification in the post-socialist city: Łódź and Leipzig two decades after socialism. *Geografie*, 120(2), 164–187.

Hołuj A (2017) Wpływ dojazdów do miasta rdzeniowego na emisję CO2 w Miejskich Obszarach Funkcjonalnych, *Studia KPZK*, 174, 362–371.

Horňáková M, Jíchová J (2020) Deciding where to live: case study of cohousing-inspired residential project in Prague. *Journal of Housing and the Built Environment*, 35, 807–827. https://doi.org/10.1007/s10901-019-09714-7

Hudec O, Džupka P (2016) Culture-led regeneration through the young generation: Košice as the European Capital of Culture. *European Urban and Regional Studies*, 23(3), 531–538.

Ivan I (2010) Advantage of carpooling in comparison with individual and public transport case study of the Czech Republic. *Geographia Technica*, 1, 36–46.

Jacobsson K (2015) *Urban Grassroots Movements in Central and Eastern Europe*. Farnham: Ashgate.

Jacobsson K, Korolczuk E (2020) Mobilizing grassroots in the city: Lessons for civil society research in Central and Eastern Europe. *International Journal of Politics, Culture, and Society*, 33(2), 125–142.

Jaczewska B, Gregorczyk A (2017) Residential Segregation at the Local Level in Poland. Case Studies for Praga Północ, Włochy and Ursynów. Miscellanea *Geographica*, 21(4), 168–178. https://doi.org/10.1515/mgrsd-2017-0032

Jakóbczyk-Gryszkiewicz J (ed.) (2015) *Procesy gentryfikacji w obszarach śródmiejskich wielkich miast na przykładzie Warszawy*, Łodzi i Gdańska. Studia KPZK PAN, CLXV, 20–42.

Jakubowicz K (1993) Musical chairs? The three public spheres in Poland. In: Dahlgren P, Sparks C (eds), *Communication and Citizenship: Journalism and the Public Sphere*. London: Routledge, 153–173.

Jałowiecki B, Łukowski W (eds.) (2007) *Gettoizacja polskiej przestrzeni miejskiej*. Warszawa: Wydawnictwo Naukowe Scholar.

Jarczewski W, Kułaczkowska A (eds) (2019) *Raport o stanie polskich miast: Rewitalizacja*. Warszawa: Instytut Rozwoju Miast i Regionów.

Jaśkiewicz M, Besta T (2014) Heart and mind in public transport: Analysis of motives, satisfaction and psychological correlates of public transportation usage in the Gdańsk–Sopot–Gdynia Tricity Agglomeration in Poland. *Transportation Research Part F*, 26, 92–101.

Jezierska K, Polanska DV (2018) Social Movements Seen as Radical Political Actors: The Case of the Polish Tenants' Movement. *Voluntas*, 29, 683–696. https://doi.org/10.1007/s11266-017-9917-2

Kaczmarek S (2001) *Rewitalizacja terenów poprzemysłowych. Nowy wymiar w rozwoju miast*. Łódź: Wydawnictwo Uniwersytetu Łódzkiego.

Kaczmarek S, Marcińczak S (2013) The blessing in disguise: urban regeneration in Poland in a neo-liberal milieu. In: Leary ME, McCarthy J (eds), *The Routledge Companion to Urban Regeneration*. London: Routledge 98–106.
Kährik A (2000) Housing Privatisation in the Transformation of the Housing System – The Case of Tartu, Estonia. *Norsk Geografisk Tidsskrift–Norwegian Journal of Geography*, 54(2), 2–11.
Kährik A, Temelová J, Kadarik K, Kubeš J (2015) What Attracts People to Inner City Areas? The Cases of Two Post-socialist Cities in Estonia and the Czech Republic. *Urban Studies*, 53(2), 1–18.
Kajdanek K (2011) *Pomiędzy miastem a wsią: suburbanizacja na przykładzie osiedli podmiejskich Wrocławia*. Kraków: Zakład Wydawniczy NOMOS.
Kajdanek K, Pluta J (2016) Aktywność lokalna w przestrzeni publicznej a potencjał grup interesu. *Przegląd Socjologiczny*, 65, 101–125.
Karolewski IP (2016) Protest and participation in post-transformation Poland: The case of the Committee for the Defense of Democracy (KOD). *Communist and Post-Communist Studies*, 49(3), 255–267. https://doi.org/10.1016/j.postcomstud.2016.06.003
Kenney P (2015) *Budowanie Polski Ludowej: robotnicy a komuniści 1945–1950*. Warszawa: Grupa Wydawnicza Foksal.
Keresztély K, Scott JW (2012) Urban regeneration in the post-socialist context: Budapest and the search for a social dimension. *European Planning Studies*, 20(7), 1111–1134. https://doi.org/10.1080/09654313.2012.67434
Killingsworth M (2007) Opposition and Dissent in Soviet Type Regimes: Civil Society and its Limitations. *Journal of Civil Society*, 3(1), 59–79.
Klimovský D, Junjan V, Nemec J (2021) Selected factors determining the adoption and use of participatory budgeting in Central and Eastern Europe. *Slovak Journal of Political Sciences*, 21(2), 230–255.
Kołoś A, Taczanowski J (2016) The feasibility of introducing light rail systems in medium-sized towns in Central Europe. *Journal of Transport Geography*, 54(2016), 400–413.
Kołoś A, Taczanowski J (2018) MożliwoŚci i dylematy rozwoju miejskiego transportu szynowego w Polsce. *Prace Komisji Geografii Komunikacji PTG*, 21(3), 31–44.
Kołsut B (2020) The import of used cars to Poland after EU accession. *Prace Komisji Geografii Przemysłu Polskiego Towarzystwa Geograficznego*, 34(2), 129–143. https://doi.org/10.24917/20801653.342.9
Komornicki T (2003) Factors of development of car ownership in Poland. *Transport Reviews*, 23(4), 413–431.
Kopeć K, Wojtowicz B (2021). Woonerf jako idea projektowania ulic zorientowana na poprawę jakości życia w mieście. *Środowisko Mieszkaniowe*, 36(36), 48–56.
Kopp A (1970) *Town and Revolution: Soviet Architecture and City Planning*, 1917–1935. New York: George Brazilier.
Kornai J (1992) *The Socialist System: The Political Economy of Communism*. Oxford: Clarendon Press.
Korolczuk E (2016) Explaining mass protests against abortion ban in Poland: the power of connective action. *Zoon Politikon*, 7, 91–113.
Kovács Z, Herfert G (2012) Development Pathways of Large Housing Estates in Post-socialist Cities: An International Comparison. *Housing Studies*, 27(3), 324–342.

Kovács Z, Wiessner R, Zischner R (2013) Urban renewal in the inner city of Budapest: Gentrification from a post-socialist perspective. *Urban Studies*, 50(1), 22–38.

Kowalczyk-Anioł J (2019) Turystyfikacja zasobów mieszkaniowych historycznych dzielnic śródmiejskich. Przykład Krakowa. *Studia Miejskie*, 35, 9–25.

Krakovský R (2018) *State and Society in Communist Czechoslovakia: Transforming the Everyday from World War II to the Fall of the Berlin Wall*. London: I.B. Tauris.

Kristiánová K (2016) Post-socialist transformations of green open spaces in large scale socialist housing estates in Slovakia. *Procedia Engineering*, 161, 1863–1867. https://doi.org/10.1016/j.proeng.2016.08.715

Królikowski JT (2020) *Poetyka miasta socjalistycznego*. Warszawa: Oficyna Wydawnicza Rewasz.

Kronenberg J, Bergier T (2010) *Wyzwania zrównoważonego rozwoju w Polsce*. Kraków: Fundacja Sendzimira.

Kronenberg J, Haase A, Łaszkiewicz E, Antal A, Baravikova A, Biernacka M, Dushkova D, Filčak R, Haase D, Ignatieva M, Khmara Y, Niţă MR, Onose DA (2020) Environmental Justice in the Context of Urban Green Space Availability, Accessibility, and Attractiveness in Postsocialist Cities. Cities, 106. https://doi.org/10.1016/j.cities.2020.102862

Kruczek Z (2018) TuryŚci vs. mieszkańcy. Wpływ nadmiernej frekwencji turystów na proces gentryfikacji miast historycznych na przykładzie Krakowa. *Turystyka Kulturowa*, 3, 29–41.

Kubeš J, Kovács Z (2020) The kaleidoscope of gentrification in post-socialist cities. *Urban Studies*, 57(13), 2591–2611. https://doi.org/10.1177/0042098019889257

Kubicki P (2011) Nowi mieszczanie – nowi aktorzy na miejskiej scenie. *Przegląd Socjologiczny*, 60(2–3), 203–227.

Kubicki P (2019) Ruchy miejskie w Polsce. *Dekada doświadczeń. Studia Socjologiczne*, 3(234), 5–30.

Kubicki P, Bakalarczyk R, Mackiewicz-Ziccardi M (2019) Protests of people with disabilities as examples of fledgling disability activism in Poland. *Canadian Journal of Disability Studies*, 8(5), 141–160. https://doi.org/10.15353/cjds.v8i5.569

Kucza-Kuczyński K, Mroczek AA (1991) *Nowe kościoły w Polsce*. Warszawa: Instytut Wydawniczy Pax.

Kwiatkowski MA (2018) Bike-sharing boom – rozwój nowych form zrównoważonego transportu w Polsce na przykładzie roweru publicznego. *Prace Komisji Geografii Komunikacji PTG*, 21(3), 60–69. https://doi.org/10.4467/2543859XPKG.18.007.10142

Lang S (2013) *NGOs, Civil Society, and the Public Sphere*. Cambridge: Cambridge University Press.

Langenohl A, Schmäing S (2020) Democratic Agnosia: The Reduction of the Political Public Sphere to Elections and Polls in Russia. *Javnost – The Public*, 27(1), 17–34. https://doi.org/10.1080/13183222.2020.1675436

Latham A, Wood PRH (2015) Inhabiting infrastructure: exploring the inter-actional spaces of urban cycling. *Environment and Planning A*, 47(2), 300–319.

Lebow K (2013) *Unfinished Utopia: Nowa Huta, Stalinism, and Polish Society, 1949–56*. Ithaca: Cornell University Press.

Lelea MA, Voiculescu S (2017) The production of emancipatory feminist spaces in a post-socialist context: organization of Ladyfest in Romania. *Gender, Place & Culture*, 24(6), 794–811.

Lenin VI (2001) Theses and report on bourgeois democracy and the dictatorship of the Proletariat. In: *Lenin VI, Democracy and Revolution. Resistance Books*, 136–149.

Light D (2004) Street Names in Bucharest, 1990–1997: Exploring the Modern Historical Geographies of Post-socialist Change. *Journal of Historical Geography*, 30(1), 154–172. https://doi.org/10.1016/S0305-7488(02)00102-0

Lipovsky C (2021) Free-floating electric scooters: representation in French mainstream media. *International Journal of Sustainable Transportation*, 15(10), 778–787. https://doi.org/10.1080/15568318.2020.1809752

Lorencová EK, Slavíková L, Emmer A, Vejchodská E, Rybová K, Vačkářová D (2021) Stakeholder engagement and institutional context features of the ecosystem-based approaches in urban adaptation planning in the Czech Republic. *Urban Forestry & Urban Greening*, 58, 126955. https://doi.org/10.1016/j.ufug.2020.126955

Lorens P (2016) Contemporary Issues in Production and Utilisation of the Common Urban Areas. *Biuletyn KPZK*, 264, 244–259.

Lowe S (2004) Overview: Too poor to move, too poor to stay. In: Fearn J (ed), *Too Poor to Stay, Too Poor to Move*. Budapest: Open Society Institute, 13–21.

Luca O, Petrescu F, Iacoboaea C, Gaman F, Aldea M, Sercaianu M (2015) Green structure in Romania: the true story. *WIT Transactions on Ecology and the Environment*, 193(12), 489–500. https://doi.org/10.2495/SDP150421

Lux M (2001) Social Housing in the Czech Republic, Poland and Slovakia. *International Journal of Housing Policy*, 2(1), 189–209.

Lux M (ed.) (2003) *Housing Policy: An End or a New Beginning?* Budapest: Local Government and Public Service Reform Initiative, Open Society Institute.

Macek J, Macková A, Kotišová J (2015) Participation or new media use first? Reconsidering the role of new media in civic practices in the Czech Republic. *Media Studies*, 6(11), 68–83.

Machon P, Dingsdale A (1989) Public Transport in a Socialist Capital City: Budapest. *Geography*, 74(2), 159–162.

Maćkiewicz B, Puente Asuero R, Pawlak K (2018) Reclaiming Urban Space: A Study of Community Gardens in Poznań. *Quaestiones Geographicae*, 37(4), 131–150. https://doi.org/10.2478/quageo-2018-0042

Mączka K, Jeran A, Matczak P, Milewicz M, Allegretti G (2021) Models of participatory budgeting. Analysis of participatory budgeting procedures in Poland. *Polish Sociological Review*, 216(4), 473–492.

Maier K (1998) Czech planning in transition: Assets and deficiencies. *International Planning Studies*, 3(3), 351–365. https://doi.org/10.1080/13563479808721719

Mailand M, Due J (2004) Social dialogue in Central and Eastern Europe: Present state and future development. *European Journal of Industrial Relations*, 10(2), 179–197. https://doi.org/10.1177/0959680104044190

Majbroda K (2018) Wrocławski Marsz Równości jako przykład karnawalizacji protestu w przestrzeni miejskiej: Perspektywa antropologiczna. *Journal of Urban Ethnology*, 16, 63–80.

Majorek A (2018) The impact of changes in methodology of participatory budgeting of Dąbrowa Górnicza on the quality of selected projects. *Prace Naukowe Uniwersytetu Ekonomicznego we Wrocławiu*, 502, 95–103.

Makowski W (2017) Jak zmieniać mobilnoŚć w dobie globalnej konkurencji miast. In: Górski R (ed.) *Odzyskajmy centra miast*. Łódź: Instytut Spraw Obywatelskich, 75–90.

Mandič S (2010) The Changing Role of Housing Assets in Post-socialist Countries. *Journal of Housing and the Built Environment*, 25, 213–226.

Marcińczak S, Musterd S, Stępniak M (2012) Where the Grass is Greener: Social Segregation in Three Major Polish Cities at the Beginning of the 21st Century. *European Urban and Regional Studies*, 19(4), 383–403.

Marcuse P (1996) Privatization and its discontents: Property rights in land and housing in Eastern Europe. In: Andrusz G, Harloe M, Szelenyi I (eds), *Cities After Socialism: Urban and Regional Change and Conflict in Post-Socialist Societies*. London: Blackwell, 119–191.

Martela B (2011) AktywnoŚć obywatelska w ramach inicjatyw nieformalnych: przyczynek do refleksji. *Acta Universitatis Lodziensis, Folia Sociologica*, 38, 105–120.

Martela B (2020) Wpływ budżetu obywatelskiego na przestrzeń polskich miast. *Urban Development Issues*, 66, 173–182. https://doi.org/10.2478/udi-2020-0021

Masik G (2012) Neoliberalizm w polityce przestrzennej dużych miast w Polsce. In: Szmytkowska S, Sagan I (eds), *Miasto w dobie neoliberalnego urbanizmu*. Gdańsk: Wydawnoctwo Uniwersytetu Gdańskiego, 211–221.

Matlovič R, Ira V, Sykora L, Szczyrba Z (2001) Procesy transformacyjne struktury przestrzennej miast postkomunistycznych (na przykładzie Pragi, Bratysławy, Ołomuńca oraz Preszowa. In: Jażdżewska I (ed.), *Miasto postsocjalistyczne – organizacja przestrzeni miejskiej i jej przemiany*, Łódź: Wydawnictwo Uniwersytetu Łódzkiego, 243–252.

Matoušek R (2013) New Municipal Housing Construction in Czechia from the Perspective of Social and Spatial Justice. *Geografie*, 118(2), 138–157.

Matynia E (2001) The Lost Treasure of Solidarity. *Social Research*, 68(4), 917–936.

Matysek-Imielińska M (2020) *Warsaw Housing Cooperative: City in Action*. Cham: Springer.

May AD (2015) Encouraging good practice in the development of Sustainable Urban Mobility Plans. *Case Studies on Transport Policy*, 3(1), 3–11. https://doi.org/10.1016/j.cstp.2014.09.001

McGarry A (2016) Pride parades and prejudice: Visibility of Roma and LGBTI communities in post-socialist Europe. *Communist and Post-Communist Studies*, 49(3), 269–277.

Mergler L, Pobłocki K, Wudarski M (2013) *Anty-Bezradnik przestrzenny: prawo do miasta w działaniu*. Warszawa: Fundacja Res Publica im. H. Krzeczkowskiego.

Mikheyev D (1987) The Soviet Mentality. *Political Psychology*, 8(4), 491–523.

Miles (2009) Public Spheres. In: Harutyunyan A, Hörschelmann K, Miles M (eds), *Public Spheres After Socialism*. Bristol: Intellect Books, 133–150.

Mokras-Grabowska J (2020) Allotment gardening in Poland – new practices and changes in recreational space. *Miscellanea Geographica*, 24(4), 245–252. https://doi.org/10.2478/mgrsd-2020-0039

Mulíček O, Seidenglanz D (2019) Public transport in Brno: From socialist to post-socialist rhythms. In: Tuvikene T, Sbignev W, Neugebauer CS (eds), *Post-Socialist Urban Infrastructures*. London: Routledge, 158–177.

Murawski M (2019) *The Palace Complex: A Stalinist Skyscraper, Capitalist Warsaw, and a City Transfixed*. Bloomington: Indiana University Press.

Murzyn MA (2006) *Kazimierz: The Central European Experience of Urban Regeneration*. Kraków: Międzynarodowe Centrum Kultury.

Nawratek K (2005) *Ideologie w przestrzeni: próby demistyfikacji*. Kraków: Universitas.

Nedović-Budić Z (2001) Adjustment of Planning Practice to the New Eastern and Central European Context. *Journal of the American Planning Association*, 67(1), 38–52. https://doi.org/10.1080/01944360108976354

Negt O, Kluge A (1993) *Public Sphere and Experience: Toward an Analysis of the Bourgeois and Proletarian Public Sphere.* Minneapolis: University of Minnesota Press.

Nikšič M (2017) Is a Walkable Place a Just Place? *The Case of Ljubljana. Built Environment*, 43(2), 214–235. https://doi.org/10.2148/benv.43.2.214

Niță MR, Badiu DL, Onose DA, Gavrilidis AA, Grădinaru SR, Năstase II, Lafortezza R (2018) Using local knowledge and sustainable transport to promote a greener city: The case of Bucharest, Romania. *Environmental Research*, 160, 331–338. https://doi.org/10.1016/j.envres.2017.10.007

Nowak M, Pluciński P (2011) Problemy ze sferą publiczną. O pożytkach z partykularnych rozstrzygnięć. In: Nowak M, Pluciński P (eds), *O miejskiej sferze publicznej. Obywatelskość i konflikty o przestrzeń.* Kraków: Korporacja Ha!art, 11–43.

Nzimande NP, Fabula S (2020) Socially sustainable urban renewal in emerging economies: A comparison of Magdolna Quarter, Budapest, Hungary and Albert Park, Durban, South Africa. *Hungarian Geographical Bulletin*, 69(4), 383–400. https://doi.org/10.15N20z1im/haunndge,eNob.Pu.lla.6n9d.4S.z4

Oglęcka E (2010) Prawne aspekty planowania terenów zieleni w miastach. *Studia Miejskie*, 2, 267–283.

Ornatowski CM (2008) Into the Breach: The Designation Speech and Expose of Tadeusz Mazowiecki and Poland's Transition from Communism. *Advances in the History of Rhetoric*, 11–12(1), 359–427. https://doi.org/10.1080/15362426.2009.10597390

Oswald I, Voronkov V (2004) The "public–private" sphere in Soviet and post-Soviet society. Perception and dynamics of "public" and "private" in contemporary Russia. *European Societies*, 6(1), 97–117.

Parysek JJ (2016) Dla kogo miasto? Dla ludzi czy dla samochodów? *Studia Miejskie*, 23, 9–27.

Perkowski P (2013) Gdańsk – miasto od nowa. Kształtowanie społeczeństwa i warunki bytowe w latach 1945–1970. Gdańsk: słowo/obraz terytoria.

Piątek G (2020) *Najlepsze miasto świata: Warszawa w odbudowie 1944–1949.* Warszawa: Wydawnictwo W.A.B.

Pichler-Milanovic N, Gutry-Korycka M, Rink D (2007) Sprawl in the post-socialist city: The changing economic and institutional context of central and eastern european cities. In: Couch C, Leontidou L, Petschel-Held G (eds), *Urban Sprawl in Europe: Landscapes*, Land-use Change and Policy, Oxford: Blackwell Publishing, 102–135.

Pickvance C (1994) Housing privatization and housing protest in the transition from state socialism-a comparative-study of Budapest and Moscow. *International Journal of Urban and Regional Research*, 18(3), 433–450.

Pieriegud J (2020) E-mobility on-demand in the Central and Eastern European countries: current trends, barriers and opportunities. *Transport Economics and Logistics*, 81, 143–154. https://doi.org/10.26881/etil.2019.81.12

Pittini A, Ghekiere L, Dijol J, Kiss I (2015) *The state of housing in the EU 2015.* Brussels: Housing Europe, the European Federation for Public, Cooperative and Social Housing.

Pixová M (2018) The Empowering Potential of Reformist Urban Activism in Czech Cities. *Voluntas*, 29, 670–682. https://doi.org/10.1007/s11266-018-0011-1

Pixová M (2020) Contested Czech cities: From urban grassroots to pro-democratic populism. *Geografie*, 118(3), 221–242.

Pixová M, Sládek J (2017) Touristification and awakening civil society in post-socialist Prague. In: Colomb C, Novy J (eds), *Protest and Resistance in the Tourist City*. London: Routledge, 73–89.

Pluciński P (2014) Miasto to nie firma! Dylematy i tożsamość polityczna miejskich ruchów społecznych we współczesnej Polsce. *Przegląd Socjologiczny*, 1, 137–170.

Plyushteva A (2019) Predictability and propinquity on the Sofia Metro: Everyday metro journeys and long-term relations of transport infrastructuring. In: Tuvikene T, Sgibnev W, Neugebauer CS (eds), *Post-Socialist Urban Infrastructures*. London: Routledge, 178–194.

Pobłocki K (2011) Prawo do miasta i ruralizacja świadomości w powojennej Polsce. In: Nowak M, Pluciński P (eds), *O miejskiej sferze publicznej. Obywatelskość i konflikty o przestrzeń*. Kraków: Korporacja Ha!art, 129–146.

Podoba J (1998) Rejecting green velvet: Transition, environment and nationalism in Slovakia. *Environmental Politics*, 7(1), 129–144.

Polanska DV (2017) Marginalizing discourses and activists' strategies in collective identity formation: The case of Polish tenants' movement. In: Jacobsson K, Korolczuk E (eds), *Civil Society Revisited: Lessons from Poland*. New York: Berghahn Books, 176–199.

Połom M, Tarkowski M, Puzdrakiewicz K (2018) Urban Transformation in the Context of Rail Transport Development: The Case of a Newly Built Railway Line in Gdańsk (Poland). *Journal of Advanced Transportation*. https://doi.org/10.1155/2018/1218041

Preoteasa I (2002) Intellectuals and the Public Sphere in Post-communist Romania: A Discourse Analytical Perspective. *Discourse & Society*, 13(2), 269–292.

Priemus H, Mandi S. (2000) Rental housing in Central and Eastern Europe as no man's land. *Journal of Housing and the Built Envirommment*, 15(3), 205–21.

Przyłęcki P (2015) Lokalne ruchy społeczne w walce o przestrzeń miejską na przykładzie Grupy Pewnych Osób. In: Maslanka T, Wiśniewski R (eds) *Kultury kontestacji. Dziedzictwo kontrkultury i nowe ruchy społecznego sprzeciwu*. Warszawa: Wydawnictwa Uniwersytetu Warszawskiego, 130–148.

Pucher J (1999) The transformation of urban transport in the Czech Republic, 1988–1998. *Transport Policy*, 6(4), 225–236.

Pucher J, Buehler R (2005) Transport Policy in Post-Communist Europe. In: Button KJ, Hensher DA (eds), *Handbook of Transport Strategy, Policy and Institutions*. Amsterdam: Elsevier, 725–743. https://doi.org/10.1108/9780080456041

Radzimski A, Gadziński J (2019) Travel Behaviour in a Post-Socialist City. *European Spatial Research and Policy*, 26(1), 43–60. https://doi.org/10.18778/1231-1952.26.1.03

Roberts K (2003) Change and Continuity in Youth Transitions in Eastern Europe: Lessons for Western Sociology. *The Sociological Review*, 51, 484–505.

Roberts P (2000) The evolution, definition and purpose of urban regeneration. In: Roberts P, Sykes H (eds) *Urban Regeneration: A Handbook*. London: Sage Publications, 9–36.

Rosik P, Stępniak M, Komornicki T (2015) The decade of the big push to roads in Poland: impact on improvement in accessibility and territorial cohesion from a policy perspective. *Transport Policy*, 37, 134–146.

Sagan I (2000) *Miasto: Scena konfliktów i współpracy*. Gdańsk: Wydawnictwo Uniwersytetu Gdańskiego.

Sagan I, Grabkowska M (2012) Urban Regeneration in Gdańsk, Poland: Local Regimes and Tensions Between Top-Down Strategies and Endogenous Renewal. *European Planning Studies*, 20(7), 1135–1154.

Sagan I, Marcińczak S (2011) The socio-spatial restructuring of Łódź, Poland. *Urban Studies*, 48(9), 1789–1809.

Sas-Bojarska A (2015) Park w metropolii – perspektywy i paradoksy. *Biuletyn KPZK*, 259, 175–193.

Scrieciu SŞ, Stringer LC (2008) The Transformation of Post-Communist Societies in Central and Eastern Europe and the Former Soviet Union: An Economic and Ecological Sustainability Perspective. *European Environment*, 18, 168–185. https://doi.org/10.1002/eet.480

Sendi R, Aalbers M, Trigueiro M (2009) Public Space in Large Housing Estates. In: Rowlands R, Musterd S, van Kempen R (eds) *Mass Housing in Europe*. London: Palgrave Macmillan, 131–156. https://doi.org/10.1057/9780230274723_6

Şerban M (2014) The Loss of Property Rights and the Construction of Legal Consciousness in Early Socialist Romania (1950–1965). *Law & Society Review*, 48(4), 773–805.

Siciarek M (2014). Narzędzia partycypacji nie gwarantują partycypacji In: Bukowiecki Ł, Obarska M, Stańczyk X (eds), *Miasto na żądanie: aktywizm, polityki miejskie, doŚwiadczenia*. Warszawa: Wydawnictwo Uniwersytetu Warszawskiego, 153–157.

Sidorenko E (2008) Which way to Poland? Re-emerging from Romantic unity. In: Myant M, Cox T (eds), *Reinventing Poland: Economic and Political Transformation and Evolving National Identity*. London: Routledge, 100–116.

Siegelbaum LH (ed.) (2011) *The Socialist Car: Automobility in the Eastern Bloc*. New York: Cornell University Press.

Siemieniako B (2017) *Reprywatyzując Polskę: Historia wielkiego przekrętu*, Warszawa: Wydawnictwo Krytyki Politycznej.

Škamlová L, Wilkaniec A, Szczepańska M, Bačík V, Hencelová P (2020) The development process and effects from the management of community gardens in two post-socialist cites: Bratislava and Poznań. *Urban Forestry and Urban Greening*, 48, 126572. https://doi.org/10.1016/j.ufug.2019.126572

Slach O, Boruta T (2012) What can cultural and creative industries do for urban development? Three stories from the post-socialist industrial city of Ostrava. *Quaestiones Geographicae*, 31(4), 99–112. https://doi.org/10.2478/v10117-012-0039-z

Smigiel Ch (2014) The production of segregated urban landscapes: A critical analysis of gated communities in Sofia. *Cities*, 36, 182–192.

Smith DM (1994) Social justice and the post-socialist city. *Urban Geography*, 15(7), 612–627. https://doi.org/10.2747/0272-3638.15.7.612

Spilková J, Vágner J (2016) The loss of land devoted to allotment gardening: The context of the contrasting pressures of urban planning, public and private interests in Prague, Czechia. *Land Use Policy*, 52, 232–239.

Springer F (2015) *13 pięter*. Wołowiec: Wydawnictwo Czarne.

Stanilov K (2007) Democracy, markets, and public space in the transitional societies of Central and Eastern Europe. In: Stanilov K (ed), *The Post-Socialist City: Urban Form and Space Transformations in Central and Eastern Europe after Socialism*. Dordrecht: Springer, 269–283.

Stenning A, Smith A, Rochovská A, Świątek D (2010) *Domesticating Neo-Liberalism: Spaces of Economic Practice and Social Reproduction in Post-Socialist Cities*. Oxford: Wiley-Blackwell.

Stevenson N (2017) E.P. Thompson and Cultural Sociology: Questions of Poetics, Capitalism and the Commons. *Cultural Sociology*, 11(1), 11–27.

Struyk RJ (1996) Housing privatization in the former soviet bloc to 1995. In: Andrusz G, Harloe M, Szelenyi I (eds), *Cities after Socialism: Urban and Regional Change and Conflict in Post-socialist Societies*. Oxford: Blackwell, 192–213.

Sükösd M (1990) From Propaganda to "OEFFENTLICHKEIT" in Eastern Europe. Four Models of Public Space under State Socialism. *PRAXIS International*, 1+2, 39–63.

Sula P (2008) Party System and Media in Poland after 1989. *Central European Journal of Communication*, 1, 145–155.

Sýkora L (2005) Gentrification in post-communist cities. In: Atkinson R, Bridge G (eds), *Gentrification in a Global Context: The New Urban Colonialism*. Abingdon: Routledge.

Sýkora L (2008) Revolutionary change, evolutionary adaptation and new path dependencies: socialism, capitalism and transformations in urban spatial organizations. In: Strubelt W, Gorzelak G (eds), *City and Region: Papers in Honour of Jiri Musil*, Farmington Hills: Budrich UniPress, 283–295.

Sýkora L (2009) New Socio-Spatial Formations: Places of Residential Segregation and Separation in Czechia. *Tijdschrift voor Economische en Sociale Geografie*, 100(4), 417–435.

Szafrańska E (2015) Transformations of Large Housing Esttaes in Central and Eastern Europe after the Collapse of Communism. *Geographia Polonica*, 88(4), 621–648. http://dx.doi.org/10.7163/GPol.0037

Szafrańska E (2016) *Wielkie osiedla mieszkaniowe w mieście postsocjalistycznym: Geneza, rozwój, przemiany, percepcja*. Łódź: Wydawnictwo Uniwersytetu Łódzkiego.

Szmytkowska M (2017) *Kreacje współczesnego miasta*. Gdańsk: Wydawnictwo Uniwersytetu Gdańskiego.

Sztompka P (1996) Trust and Emerging Democracy: Lessons from Poland. *International Sociology*, 11(1), 37–62. https://doi.org/10.1177/026858096011001004

Szynkarczuk A (2015) Behind the Iron Gate Housing Estate: Theory and Praxis of Green-Space Creation Before and After the End of Modern Movement. *Architeruae et Artibus*, 7(4), 54–64.

Taczanowski J, KołoŚ A, Gwosdz K, Domański B, Guzik R (2018) The development of low-emission public urban transport in Poland. *Bulletin of Geography. Socio-economic Series*, 41(41), 79–92. http://doi.org/10.2478/bog-2018-0027

Tammaru T, Marcińczak S, van Ham M, Musterd S (2015) *Socio-Economic Segregation in European Capital Cities: East Meets West*. London: Routledge.

Tarkowski J (1994) *Władza i społeczeństwo w systemie autorytarnym*. Warszawa: Instytut Studiów Politycznych.

Temelová J (2007) Flagship developments and the physical upgrading of the post-socialist inner city: The golden angel project in Prague. *Geografiska Annaler: Series B, Human Geography*, 89(2), 169–181.

Tischner J (1992) *Etyka solidarności i Homo sovieticus*. Kraków: Społeczny Instytut Wydawniczy Znak.

Todorova M, Gille Z (eds.) (2010) *Post-communist Nostalgia*. New York: Berghahn Books.
Tonkiss F (2013) *Cities by Design: The Social Life of Urban Form*. Cambridge: Polity Press.
Tóth A, Duží B, Vávra J, Supuka J, Bihuňová M, Halajová D, Martinát S, Nováková E (2018) Changing Patterns of Allotment Gardening in the Czech Republic and Slovakia. *Nature and Culture*, 13(1), 161–188. https://doi.org/10.3167/nc.2018.130108
Trammer K (2019) *Ostre cięcie: Jak niszczono polską kolej*. Warszawa: Wydawnictwo Krytyki Politycznej.
Trendov, NM (2018) Comparative study on the motivations that drive urban community gardens in Central Eastern Europe. *Annals of Agrarian Science*, 16(1), 85–89.
Tsenkova S (2009) *Housing Policy Reforms in Post Socialist Europe: Lost in Transition*. Heidelberg: Physica-Verlag.
Turovets M (2020) When Public is Not Enough: A Local Russian Social Movement between Suppression and Depoliticisation. *Javnost - The Public*, 27(1), 35–47. https://doi.org/10.1080/13183222.2020.1675426
Tuvikene T (2016) Strategies for Comparative Urbanism: Post-socialism as a De-territorialized Concept. *International Journal of Urban and Regional Research*, 40(1): 132–146.
Tuvikene T, Sgibnev W, Neugebauer CS (2019) Introduction: Linking post-socialist and urban infrastructures. In: Tuvikene T, Sgibnev W, Neugebauer CS (eds), *Post-Socialist Urban Infrastructures*. London: Routledge, 1–20.
Twardoch A (2019) *System do mieszkania: Perspektywy rozwoju dostępnego budownictwa mieszkaniowego*. Warszawa: Fundacja Nowej Kultury Bęc Zmiana.
Udvarhelyi É (2009) Reclaiming the streets: Redefining democracy. *Hungarian Studies*, 23(1), 121–145.
Utekhin I (2018) Suspicion and mistrust in neighbour relations. A legacy of the soviet mentality? In: Mühlfried F (ed), Mistrust: Ethnographic Approximations. Bielefeld: transcript Verlag, 201–218.
Ważyk A (1950) *Nowy wybór wierszy*. Warszawa: Spółdzielnia Wydawniczo-Oświatowa “Czytelnik”.
Węcławowicz G (1998) Social polarisation in post-socialist cities: Budapest, Prague and Warsaw. In: Enyedi G (ed), *Social Change and Urban Restructuring in Central Europe*. Budapest: Akadémiai Kiadó, 55–66.
Węgleński J (1992) *Urbanizacja bez modernizacji*. Warszawa: Państwowe Wydawnictwo Naukowe.
Winkler K (1952) Czerwony autobus. Retrieved from: http://bibliotekapiosenki.pl/utwory/Czerwony_autobus_(sl_Kazimierz_Winkler)/tekst
Witkowska A, Witkowski A (2006) Pierwszy, jedyny, najlepszy. https://nomus.gda.pl/pl/zbiory/w/anna-i-adam-witkowscy/pierwszy-jedyny-najlepszy
Zarecor KE (2011) *Manufacturing a Socialist Modernity: Housing in Czechoslovakia, 1945–1960*. Pittsburgh: University of Pittsburgh Press.
Zarecor KE (2013) Infrastructural thinking: Urban housing in former Czechoslovakia from the Stalin era to EU accession. In: Murphy E, Hourani NB (eds), *The Housing Question: Tensions, Continuities, and Contingencies in the Modern City*. Aldershot: Ashgate, 57–78.
Zarecor KE (2018) What Was so Socialist about the Socialist City? Second World Urbanity in Europe. *Journal of Urban History*, 44(1), 95–117.

Zaremba M (2012) *Wielka trwoga: Polska 1944–1947: ludowa reakcja na kryzys*. Kraków: Wydawnictwo Znak.
Zdyb M (2017) Proces rewitalizacji a jakość życia mieszkańców–projekty Zielone Polesie i woonerfy w Łodzi. *Space–Society–Economy*, 21, 73–97.
Žídek L (2019) *Centrally Planned Economies: Theory and Practice in Socialist Czechoslovakia*. Abingdon: Routledge.
Zinoviev A (1986) *Homo Sovieticus*. London: Paladin Books.

Part II

Commoning the post-socialist city

Evidence from Poland

3 Towards the city as a commons

The changing public discourse in Poland between 1989 and 2019

Discourse analysis as a key to understanding urban change in Poland after socialism

When referring to how socialism dawned upon Central and Eastern Europe, the term revolution is commonly used to denote the abrupt, and violent character of the process. However, it was a process indeed, one which had begun long before the Soviet Banner of Victory was planted on the roof of Reichstag in the spring of 1945. It is obvious that revolutions do not just appear out of nowhere but are preceded by years, if not decades, of minor shifts and adjustments which add up to eventual sweeping change. Although separate signals paving its way may be easy to ignore or misinterpret, they tend to form patterns with the benefit of hindsight, when they are investigated together and systematically throughout an extent of time. As noted by Rebecca Solnit (2016: xv), 'uprisings and revolutions are often considered to be spontaneous, but less visible long-term organizing and groundwork—or underground work—often laid the foundation'. The same author brings into focus the role of less apparent actors than insurgents and revolutionists themselves:

> Changes in ideas and values also result from work done by writers, scholars, public intellectuals, social activists, and participants in social media. It seems insignificant or peripheral until very different outcomes emerge from transformed assumptions about who and what matters, who should be heard and believed, who has rights.
>
> (2016: xv–xvi)

The communicative aspect of public discourse is also highlighted by G. Thomas Goodnight (1987: 429), who claims that 'while [it] makes open and common collective preference, it also provides an arena where interests conduct controversy and openly struggle for power'. A close study of changing public discourse helps to apprehend the substance and underpinnings of any socio-cultural transformation. Discourse analysis, therefore, offers important insight into the process of redefinition of the urban common good in Poland after 1989, as reflected in the shift from the paradigm of the city as a communal infrastructure to subsequent visions of the city as a commodity and as a commons.

DOI: 10.4324/9781003089766-6

The transition period in Poland was a time of fundamental changes which happened simultaneously within multiple orders: the political, the social, the economic, the spatial, and so on. Some such changes were immediate, others proceeded slowly. All of them imprinted on the public sphere, the prevailing discourses, narratives, and values. But old connotations did not necessarily fade away with the development of new understandings. Shortly after 1989, the concept of common good was—at the same time—semantically overcharged, and devoid of meaning. On the one hand, it carried considerable ideological weight acquired over 40 years of applied socialism. On the other hand, the loss of credibility, and the resulting fall of the communist regime, had sapped its essence by the end of the 20th century.

On account of the current comeback of the common good to the public realm, it is interesting to observe the process of renegotiation of its content. It must be underlined that in Polish the term 'common good' is used to denote general welfare and a shared resource. The lack of a separate word, equivalent to the English 'commons', adds an additional layer of interpretation to the ongoing career of the paradigm of the city as a commons, which seems to draw on both ideas at once. Nevertheless, it is the first meaning which has reigned, at first being closely linked to the issue of low social capital. As stated by the authors of a longitudinal social study, Poland is 'a country of increasingly effective individuals and a continuously ineffective community' (Czapiński 2015: 362). Subsequent editions of the study have proven that, between 2007 and 2011, around half of respondents were indifferent to various forms of social behaviour undermining the common good, while only 15.6 per cent engaged in activities for the benefit of the local community (Czapiński and Sułek 2011).

The challenges faced after socialism by the troubled condition of urban commonality in Polish cities appear in their full complexity when a broader socio-political context is taken into consideration. As elsewhere in CEE, urban development in Poland since 1989 has been strongly affected by both the weakness of the young democracy and the forcefulness of neoliberalism (Sagan 2017). Looking back at Poland at the turn of the 1990s, it is important not to underestimate the contemporaneous high regard for the long-yearned-for political and economic freedom. Since democracy was commonly regarded 'as the "natural" political accompaniment of capitalism' (Žižek 2009: 132), the expectation was that the latter would infallibly prompt the former. Thus, economic transition was placed at the centre of transformation, while issues of citizen participation and quality of governance, especially at the local level, were largely neglected. While even the Democratic Left Alliance, the major post-communist party ruling in the 1990s and early 2000s, took to the neoliberal vision, it was during the two-term government of the liberal centrist Civic Platform—between 2007 and 2015—that the already settled narratives of individual responsibility, entrepreneurialism, and trickle-down effect, gained momentum.

With the public debate promoting the symptomatic treatment of the costs of transformation over a more systematic remedy through a 'socially responsible model of the state' (Koczanowicz 2016: 82), the public sphere maintained its neoliberal bias, and the collective perspective of social solidarity was ultimately left behind. The ensuing abandonment of the disadvantaged reinforced the effects of the growing social divide between 'two Polands'.[1] Social discontent was next channelled by the national conservative party, Law and Justice, campaigning for the parliamentary elections of 2015 under the banner of *dobra zmiana* (good change), and earned them a landslide victory. Although the populist slogan was grounded in the rhetoric of the common good (Łabendowicz 2018), the trend of social fragmentation was hardly reversed in the coming years.

The consequences of the overall political unsteadiness, and the overlooked social agenda at the central level for cities, have been twofold. On the one hand, constant government reshuffles,[2] a series of corruption scandals and the neglect of social policies added to the sense of the ineffectiveness of the public sector, commonly methaphorised as 'the cardboard state' (Czech and Kassner 2021: 141). On the other one, the unrestricted power of capital helped the economic elites to seize the opportunity and assume a privileged position in the reshuffling urban arenas. Cities were especially susceptible to the interplay of these two factors under the hardships of the transformation processes. In addition, the model of local governance, built almost from scratch, did not take into account sufficient participation of local citizens. Just like the disregard of the social dimension by the state contributed to the increase of support for populist parties, resistance against the neoliberal agenda and citizen disempowerment at the local level gave way to a rise of urban activism under the banner of the right to the city.

To fully grasp the unfolding narrative of the urban common good, the temporal scope adopted in this chapter covers three decades, from 8 March 1990 to 11 March 2020. As for the opening date, it stands for the passing of the Act on Commune Self-Government by the new democratic parliament. Inaugurating the process of decentralisation of power, it was critical to the restoration of the political autonomy of Polish cities and towns. The day closing the timeframe was marked by the World Health Organization officially announcing the COVID-19 pandemic. As signalled in the Introduction, forasmuch as the unexpected global spread of the SARS-CoV-2 virus impacted all aspects of urban life, providing an adequate caesura for the new era for cities worldwide, it may also be seen as a farewell to the optics of post-socialism in CEE (cf. Müller 2019).

Having explained the focus and timespan of the study, next comes the specification of the discursive channels selected for the analysis. Drawing on a classification based on the institutional domain (Witosz 2016), this chapter investigates transformations within the following types of discourses: legal, media, and academic. Representatives of these three categories—legislators, media commentators, and experts—have been identified by sociologist Anna

Giza (2013b) as key 'broadcasters' of content in the public sphere, or 'symbolic intermediaries', whose actions and opinions significantly shape the public discourse. While the three channels set out to inform the mainstream public sphere, each of them follows its own rules. Legal discourse, comprising both legal narrative and legal interpretation (Cheng and Danesi 2019), is mostly normative in character. Media and academic discourses are, to varying degrees, more descriptive, analytical, and/or critical. Reliability of the former depends on the political and ideological stance of a particular media outlet while the expected impartiality of the latter is safeguarded, at least theoretically, by scientific rigour.

The order of this analysis roughly follows a general sequence in which the three interrelated domains might have cross-influenced each other over the research period. The **legal and administrative sphere** provided the official framework for the new interpretation of 'the common' after 1989. Therefore, the legal discourse is analysed first. However, the earliest signs of the ongoing changes in the public sphere are typically first reflected by **the media**, since they are the quickest to detect and react to any new trends in public perceptions and actions, as well as because of their agenda-setting influence (McCombs and Shaw 1972). These changes are next analysed and discussed in **academic research**. The results of dissemination of new ideas and rationales are likely to be absorbed and incorporated by the legal discourse—provided that the resulting regulations feed the public discourse, the whole cycle starts again.

The choice of the corpus in each of the three types of discourse was informed by the functions and purposes of relevant texts circulating in the public sphere. Starting with legal documents, such as statutory law and administrative acts issued at both national and local levels, the inquiry tends toward content analysis whose primary focus concerns recognition and prioritisation of the 'common' in relation to urban space and shared urban amenities. The top-down provenance of the analysed sources offers insight into their hegemonic influence within the public sphere, as well as investigates the potential for raising opposition essential for the emergence of the counterpublics. In comparison with the legal domain, the media usually reach wider audiences. Among the variety of media outlets available today, it is the digital media which have become the main transmitter of novel urban ideas and narratives (McQuire 2008, Grabkowska et al. 2013). However, traditional print media in Poland continue to be more valued in terms of their perceived accessibility, reliability, and opinion-forming potential. The results of a study conducted by Bierwiaczonek et al. (2017: 177) among citizens in three Polish cities—Gdańsk, Gliwice and Wrocław—demonstrate that print issues of newspapers are recognised as the most popular arenas for debating urban issues (75.8 per cent of indications).[3] In line with this finding, the body of texts selected for the purpose of my study is limited to electronic versions of newspaper and magazine articles which had first appeared in print. Scholarly literature selected for this particular study includes texts relating to urban common good. Some of the works have been already cited in Chapter 2 to document the shift of discourse, from 'city as a commodity' to 'city as a commons'. Here, they are summoned again to serve a more specific purpose. Namely, to provide a basis for deconstruction of the urban common good as an academic concept transpiring to other spheres of the public domain.

Occasional and unassuming: legal notions of urban common good

Albeit gradually discredited during the socialist period, the idea of common good in Poland was not entirely abandoned in 1989. In the same speech that introduced a policy of the thick line (see Chapter 2), Tadeusz Mazowiecki referred to common good as the foundation for the incoming legal system. Namely, he identified citizen participation and sense of social justice as prerequisites for a democratic, law-abiding state and went on to say that 'only the law which aims at the common good can enjoy respect and social authority' (*Chcę być…* 1989). However, Polish legislation has so far failed to define it in general, not to mention the urban, context. The following overview includes legal documents which come closest to filling this void. While I investigate the use of the notion of 'common good', I also seek specific variants of the term, less directly relating to the conceptualisation of urban common good laid out in Chapter 1. To adequately trace the evolution of the legal discourse touching upon urban common good after 1989, the relevant documents are discussed in chronological order.

Understandably, the framing of the new legal order in Poland did not happen overnight. Due to the enormity of the task, as well as because of frequent government reshuffles in the early years of transition, it took almost eight years to draft a new constitution. Similarly delayed were any acts of law critical for setting up a new urban agenda/programme. One of the earliest legislative changes concerned the restitution of the local government by the **Act on Commune Self-Government of 8 March 1990**. Guided by the principle of subsidiarity, it was the first step towards decentralisation, devolving much of the power to the lowest administrative level of *gminas* (communes).[4] Yet, despite the fact that the opening article of the Act defined gmina as 'a self-governing community and the relevant territory', the empowerment of local inhabitants was purely formal, as the remainder of the document confined their influence on decision-making to electing communal authorities. The sole opportunity for any direct citizen control was provided by the rather occasional instance of referendum voting, usually limited to decisions on dismissing municipal authorities, or self-taxation for public purposes. Availability of other forms of civic participation, such as public consultations, remained quite restricted and their results devoid of binding force. Until an amendment to the Act was introduced in 2018, the local communities had not even been entitled to initiate resolutions—unless granted this right, via a complicated procedure, by the city council.

In comparison with the socialist era, the post-socialist model of representative democracy at the local level was a major leap forward—back in the early 1990s it fulfilled the conditions for democratic legitimisation and social acceptance of the local authority (Kasiński 2018: 12). Nevertheless, from today's perspective, the full transfer of responsibility for the common good from citizens to their representatives imprinted the whole transformation period. To this day, the input of city dwellers in local decision-making in Poland is rather minimal, which affects the urban common good in numerous ways, and goes against the provisions of the European Charter of Local Self-Government, adopted in Poland in April 1993.

Just like its predecessors, passed in 1961 and 1984, the **Act on Spatial Development of 7 July 1994** contains no direct reference to the common good (Jędraszko 2005). Instead, it prioritises the right to private property, listed among the conditions which should be taken under consideration whenever conflicts between interests of citizens, self-governing communities, and the state, arise.[5] Property rights are also promoted in Article 3, which, on the one hand, guarantees the right to development of land to anyone holding a property title, and, on the other one, ensures protection of one's own legal interest arising from a property right whenever lands owned by other people or institutions are developed. Regardless of the rationality of both provisions, their prominent position within the document is quite significant. Apart from a formal recognition of the 'sacred right to property', the Act remodelled the approach to space management. The title of the document severed it from both the Spatial Planning Act of 1984, and from the system of central planning as a rule. Whether intentional or not, such renaming epitomises how in the 1990s the importance of comprehensive spatial planning lessened as the focus switched to the development of space. While the Act of 1994 reserved the status of local law to local zoning plans, local planning authorities were no longer legally obliged to adopt them.

A similar attention shift from (central) urban planning to (local) development of urban space in keeping with private interests is also manifest in the evolution of regulations within the **Construction Law** (1994). Article 4 of the original document required that the architectural form of any built structure, and related equipment, conform to the surrounding landscape/cityscape, and buildings. The provision has since been amended, and its current version secures the owner's elementary right to land development as long as the construction project complies with the currently applicable regulations. The replacement of the concern for the quality of spatial planning with a principle of freedom of construction (Asman and Niewiadomski 2018) reflects a culmination of privatism and commodification of urban space typical of the transition period.

Explicit reference to common good had not appeared in the Polish legal discourse until the **Constitution of the Republic of Poland** was finally adopted in 1997.[6] The common good is invoked as early as in Article 1—by proclaiming the Republic of Poland as 'the common good of all its citizens', it clearly breaks with the proletarian narrative of the constitution of the Polish People's Republic (1952), according to which 'the power was vested in the working people of town and country' (Art. 1.2). Common good is next brought up in Article 82, which establishes that cherishing it is a citizen's duty. Interestingly, throughout the lengthy drafting of the new Constitution, different understandings of 'the common good' divided the legislators. Marek Piechowiak (2012: 172) recounts this as a clash of two constitutional paradigms. The first one originates from the classical reflection on common good, human rights and the priority of the individual within a political community (*dobro wspólne*). It, therefore, relates to the first Polish constitution, Governance Act of 3 May (1791). Contrastingly, the other approach is rooted in the étatist tradition and focuses on the state as a common responsibility (*wspólne dobro*). Here, both the word order and the interpretation echo the opening article from the April Constitution of 1935. It was the latter version which had been more popular at first. However, out of the many bills proposed over the years, the document which was eventually passed uses the form *dobro wspólne*.[7] Notwithstanding the attempts to differentiate between the two variants, they have been treated as synonyms in Polish legal practice. For instance, the case law of the Constitutional Tribunal tends to use them interchangeably (Piechowiak 2012). This confusion results from the fact that regardless of the prominent position of the common good in the Constitution, the term lacks any specific explanation. Consequently, its application, not least in spatial planning, has been problematic (Jędraszko 2005).

Following the parliamentary elections of 1997, the self-government reform continued by means of reorganisation of the country's administrative division into a three-tier system. With effect from 1st January 1999, the number of voivodeships—or provinces—was reduced from 49 to 16, and an intermediate level of powiats (counties) was added between them and gminas, or communes (Act on Powiat Self-Government 1998, Act on Voivodeship Self-Government 1998, Act on the Introduction... 1998). While powiats typically consist of gminas, major cities, for example, those with a population over 100,000 inhabitants, and the former seats of voivodeships—hold the idiosyncratic status of the city with powiat rights (*miasta na prawach powiatu*). Technically, they remain gminas which additionally perform the supralocal tasks of powiats. The resulting emancipation of larger Polish cities was given further momentum by the **Act on the Direct Election of a Commune Head, Town Mayor and President** (2002), which reinforced the authority of town and city mayors. From then on, they would be elected not from among, and by, the councillors but in 'general, equal and direct elections, by secret ballot' (Art. 2), which heralded a fundamental change in the system of power, thus disrupting the previous relative balance between legislative and executive bodies within the communes (Kasiński 2018). As evidenced by Adam Gendźwiłł and Paweł Swianiewicz (2017), such a modification indeed gave way to the new breed of 'strong mayors'—long-lasting incumbents and non-partisan leaders, independent not only from the local councillors but also from political parties at the national level.

The same Act amended the Act on Commune Self-Government of 1990 by mandating the councillors 'to serve the good of the self-governing local community', and supplementing the wording of the oath of office with a declaration that their duties would be performed 'in cognizance of "the good of [the] commune and its inhabitants"' (Art. 23). Correspondingly, under Article 29, the mayor pledges to obey the law and to hold the office 'for the sake of public good and citizen welfare'. Since all three regulations are rather declarative, their addition remains without much bearing on reality as neither the councillors nor the mayor are bound by constituents' instructions, with autonomy throughout a four-year term in office largely unlimited.

In 2003, the **Act on Spatial Planning and Development** replaced the act of 1994. The resurgence of the term 'planning' in the very title of the statute would suggest a readjustment to spatial policy, but the change was rather nominal. Citizens were assigned only a slightly more active role in shaping the space in their vicinity—new regulations afforded local communities the right to participate in public discussions on draft local zoning plans, and to submit comments (Art. 17). Much as it was a necessary modification, it failed to spur participation in spatial planning. Citizens were still excluded from drafts' development—exactly the stage at which the principles of managing urban space as a common resource could be effectively negotiated. More importantly, the Act repealed any local zoning plans enacted before 1995. From then on, in the absence of a local plan in force, arbitrary decisions on building conditions and land development could be issued instead (Art. 2). Designed as an ad-hoc measure, and relying on discretionary grounds, the instrument not only remained beyond citizen control but also, due to its subjectivity, facilitated favouring individual interests, and, at worst, enabled corruption and abuse of authority. By 2015, over half of construction permits in Poland were based on such decisions (National Spatial Development Concept 2030 2012: 163), which further eroded spatial order, and—as such—violated the constitutional principle of common good.

While the term 'common good' is absent from the Act, the document gives some attention to public interest, featured in the glossary (Art. 2.4) as a 'generalised goal of aspirations and activities which meet the objectified needs of the general public or local communities related to spatial development'. Such a definition echoes Asen's (2017) outlining of public (common good) but in a much narrower scope. Moreover, aside from the glossary, 'public interest' only occurs twice throughout the original version of the Act. First, it closes the list of basic considerations in spatial planning and management (Art. 1.2), where it is preceded by spatial order, architectural and landscape values, natural environment, cultural heritage, health and safety, economic value of space, state defence and security, and, significantly, the right to property. Next, it reappears in Article 6.2 as a potential limitation to development of land to which one has legal title. Although indirectly, the whole set of regulations passed in the Act put the right to private property and individual interest before collective public interest, and common good.

As of 1 May 2004, over a year after the adoption of the Act on Spatial Planning and Development, Poland became a member of the European Union. This required adjustment of domestic law and public policies to the *acquis communautaire*. Moreover, any new EU legislation would immediately apply to Poland, giving the Community law potential to influence public discourse in Poland. Nevertheless, the meaning of common good employed in EU legal acts has been—just like in the Polish legislation—rather vague, and so far its potential has remained unfulfilled (Żurawik 2014).

Following the parliamentary elections of 2005, the cabinet of Jarosław Kaczyński finally addressed the issue of housing. The programme *Rodzina na swoim* (Family on their own) offered financial assistance in monthly mortgage repayments to young married couples and single parents who were getting on the property ladder. It is estimated that between 2007 and 2012, that is, the time scope of the programme, the number of beneficiaries—mostly from large cities—reached over 180,000 (Radzimski 2014). In 2014, the coalition of Civic Platform and Polish People's Party implemented a new impression of the programme, named *Mieszkanie dla młodych* (Housing for the young). This time, state aid facilitated the purchase of a newly built flat and took the form of a one-off subsidy to the private contribution for a mortgage.[8] What linked the two programmes was their targeted support of private ownership and newly built housing. Therefore, both have been criticised for limiting the role of the public and cooperative housing sector in favour of private developers, reinforcing dependence of the housing market on banking institutions, inflating housing prices, and bolstering any negative consequences engendered by progressive suburbanisation (Drozda 2016: 57). Even from the semantic point of view, the names of the programmes are tellingly dissociated from the notion of common good, hinting at a privatist orientation, and a lack of political will to provide a systemic solution to the housing issue.

Returning to the chronological order of events, 2007 marked the beginning of the longest period of uninterrupted government in post-socialist Poland—the eight-year coalition of the Civic Platform and Polish People's Party. As a time of relative stability, it saw a breakthrough in the right to the city narrative advocated by rising urban movements. May 2011, in the wake of a local grassroots initiative, saw a pioneering participatory budget at the city level, implemented in Sopot. A month later, Poznań hosted the Congress of Urban Movements, with almost a hundred participants from over a dozen Polish cities in attendance. Since then, participatory budgeting as such has been taking root countrywide, reaching a third of all municipalities in 2016 (Pistelok and Martela 2019), and four more editions of the Congress took place in cities across Poland. The political impact of both events was felt in the aftermath of the local elections of 2014—even though only a small share of candidates with activist background had won seats on city councils, many of their postulates were acknowledged and incorporated into local urban policies.

The urban agenda on the central level was not affected straight away. In 2012, **the National Spatial Development Concept 2030** was adopted, featuring a polycentric network of large competitive and innovative urban centres, as well as integration and cohesion of space, among its key objectives, anchored in the guiding principle of sustainable development.[9] While the document acknowledges the inadequacy of the existing legal solutions concerning public participation in planning at the local level (164), and advocates 'full and real openness of planning and enabling active public participation at all planning stages' (174), it also refrains from any specific recommendations. The outlined measures to be deployed in pursuit of restoration and consolidation of spatial order are far from concrete and depart from the idea of common good in terms of agency. Gathered under the umbrella term of 'public interest protection', they include, for example, 'counteracting appropriation of public space and taking care of public space', 'protecting quality and identity of natural and urbanised landscape', and 'ensuring appropriate living conditions to rural and urban populations' (164–165).

The signs of a slight discourse redirection appear in the long-term strategy of national development adopted in 2013, in which 'citizens cooperating and engaging in collective action for the common good' account for the vision of Poland in 2030 (*Polska 2030* 2013: 122). Likewise, common good is the centrepiece of a ministerial report published a year later, titled 'Common Space, Common Good: Best Practices in Shaping Spatial Order', and featuring a compilation of twenty-eight exemplary spatial transformations[10] (Gontarz and Gutowska 2014). The idea of the common good is presented there as 'a foundation of existence of cities and a new paradigm of regional development', associated with the 'capacity for dialogue and collective action', and with 'responsibility for another and aiming at achieving common benefits' (118). Urban public transport, public space, and communal housing resources are recalled as examples of 'material expressions' of the common good. Furthermore, the document addresses the matter of disregard for common good, proposing public consultations engaging the local communities, and educational workshops, as efficient remedies.

The year 2015 brought three long-awaited regulations on pressing urban issues. The so-called **Landscape Act** (Act on the Amendment… 2015), passed in April, introduced a series of legal instruments to protect the cityscape. Provisions such as the possibility to ban fences and outdoor advertising boards (Art. 7.5) potentially imposed limitations on the right to private property instead of extending it, which may be viewed as a certain breakthrough. The Act amends other legislation mainly by adding new or altering existing regulations. As such, it does not operate with the concept of common good, but several references can be found between the lines. For instance, Article 8 updated the legal understanding of cultural landscape featured in the Act on Protection and Care of Monuments (2003) to involve the components of Asens's (2017) common good—the relationship between people's perceptions and actions with the space they inhabit.

On a similar note, the **Act on Revitalisation** (2015), adopted in early October, countervailed the previous neglect of the social dimension in revitalisation undertakings (see Chapter 2). The prerequisite of inclusion of local communities as key stakeholders and beneficiaries at all stages of revitalisation processes entered the legal framework of public participation in Poland to an unprecedented level. In addition, Article 41 of the Act significantly amended the Act on Spatial Planning and Development (2003) to highlight the aspect of commonality. Firstly, by expanding the catalogue of conditions prevalent in spatial planning and management with, inter alia, ensuring public participation in planning processes and maintaining openness and transparency of planning procedures (2003, Art. 1.2). Secondly, by compelling the local authorities to 'weigh the public interest and private interests' in decision-making (2003, Art. 1.3).

Lastly, in late October 2015—just before Law and Justice won an absolute majority in the parliamentary elections—the council of ministers approved the resolution on **National Urban Policy 2023** (2015). Intended as a compass for public administration in the field of urban strategy at the national level, it represents a clear continuation of the adopted strategy of sustainable development with an additional focus on life quality. Aiming at Polish cities to become 'liveable cities', the model covers ten thematic threads: shaping urban space, public participation, transport and urban mobility, low-carbon and energy efficiency, urban regeneration, investment policy, economic development, environmental protection and adaptation to climate change, demographics, and urban area management. Common good comes up as one of the priorities in developing public participation—'building a shared sense of responsibility for the common good' is set out as a 'fundamental principle' enabling city management 'from a perspective wider than merely that of individual interests' (34). However, the suggested set of tools for urban governance—for example, participatory budgeting, citizens' initiative to propose bylaws, on-line consultations, workshops, questionnaire surveys, or study tours—amounts to nothing more than broad consultation. A deeper understanding of common good is only hinted at in the section discussing—in a rather idealistic manner—the quality of urban life:

> in the process of creating a liveable city, with open and attractive public spaces, the feeling of estrangement, alienation and lack of community disappears. Opportunities arise instead—of meetings and of interplay between different groups of users and different lifestyles. 'Common space' not only fosters the ability to identify with the place, but also to cooperate and share responsibility for it.
>
> (18)

In 2016 a new government, led by Law and Justice, issued the National Housing Programme (*Narodowy Program Mieszkaniowy*), thus symbolically breaking with the existing policy targeted at the young and creditworthy. Considering housing as one of the basic social needs, it promised to improve accessibility of flats for the previously disregarded social groups, such as the elderly, and low-income households. At the same time, it unveiled a wide spectrum of instruments—from declared support for bottom-up cooperatives, through the governmental package on affordable housing (*Mieszkanie+*), to amending regulations facilitating commercial development. By embracing the socially just approach to the housing issue, the programme recalled its position as a common resource and broke off from the narrative of 'sacred property'.

The latter was, however, prioritised again in the short-lived Act on the Amendment of the Act on Nature Conservation (2016), dubbed 'lex Szyszko' after the incumbent minister of environment. Relaxing regulations concerning felling trees in private real estates, it relieved the owners from the prior obligation to obtain a permit from the local authorities. Following a wave of protests, the regulations were lifted after several months, with irrevocable damage to green urban infrastructure already a new reality—in Łódź, the amount of tree canopy lost during the policy liberalisation period accounted for 40 per cent of the total canopy area removed within the prior decade (Kronenberg et al. 2021).

The **Act on Amenment of Certain Laws to Increase Citizens' Participation in the Process of Election, Operation and Control of Some Public Bodies of 11 January 2018** (2018) may be interpreted as another nod to the common good in terms of governance—as is evident from the title, it aimed to increase the scope of public participation on a host of government levels. Among other regulations, the amendments to the Act on Commune Self-Government (1990) established participatory budgeting, a 'special form of public consultation' (Art. 1.1), introduced a requirement for live streaming of council deliberations, and dissemination of such recordings (Art. 1.6), as well as awarded local communities the right to initiate bylaws (Art. 1.13). The Act also introduced obligatory participatory budgeting in cities with powiat rights, capping the amount to be spent at 0.5 per cent of the annual municipal expenditure. However, these provisions were rather symbolic since all Polish city counties had already implemented the tool and the financial criterion had not been met in a few of those (Mączka et al. 2021: 480). Moreover, in line with Law and Justice attempts at recentralisation (Wojnicki 2020), some other amendments introduced by the Act changed the electoral rulebook to favour party candidates and reinforce particracy at the local level (Kasiński 2018).

The presented overview reveals that the notion of common good was rather uncommon in Polish legal discourse between 1990 and 2019. Regardless of its prominent position in the constitution, and occasional appearances in regulations on local governance, housing, and urban regeneration—especially towards the end of the analysed period—I found it to be missing from legislation concerning other urban issues discussed in Chapter 2: public transport, green infrastructure, or public space. No reference to common good or public interest can be found in the Act on Nature Conservation (2004), nor in the Act on collective Public Transport (2010), even though both operate with the notion of sustainability. The frail legal footing of the commonality of urban space in the existing legal acts mobilised the community of Polish urban planners to meet in 2004 in Gdańsk at a Congress of Polish Urban Planning and debate under the slogan 'The city—a common good and collective responsibility'. Five years later, participants of the third edition of the Congress, held in Poznań, adopted Public Space Charter (2009) to counteract the disregard for public space and call for its recognition as a common resource.

Understandably, legal acts and administrative regulations only set certain standards without a full guarantee of end results of their implementation—how they translate into praxis depends on many factors. However, the absence of common good from the analysed documents is significant. It is also hardly surprising given that the constitutional accentuation of this principle is rather symbolic. The evident primacy of private interest related to property rights and insufficient citizen participation in urban planning in the legislative framework add to the overall inclination of the legal discourse towards a paradigm of the city as a commodity rather than that of a commons. Yet, the caesura of 2011 seems to have had, if only a slight, impact on subsequent legislation. This butterfly effect, which was even more felt in print media discourse, points to the significance of grassroots mobilisation in the shaping of public discourse in Poland.

The unravelling of urban common good in print media

Mass media, hailed as the fourth estate in the modern concept of separation of powers, not only reflect the public sphere but also shape it, introducing new paradigms, and undermining the prevailing ones (Peterson et al. 1966). The latter feature gains additional weight in the post-socialist context, where independent journalism has resurfaced after decades of state control and censorship. Although media systems in CEE vary, similar processes have formed and shaped them (Dobek-Ostrowska 2019). These include, for example, rapid deregulation of the press industry, and its internationalisation due to increased involvement of foreign capital following 1989. The key transformations of the Polish mediascape have included pluralisation of information and communication channels, as well as the more recent trends of tabloidisation and digitalisation of the media sphere. Commercial broadcasters dominated the media sector relatively soon, yet since the mid-2010s media freedom has been increasingly impinged upon by such practices of the Law and Justice government as withdrawal of state advertising from unloyal commercial media (*Democracy Declining* 2021).

Despite political pressures on independent journalism, the commercial media in Poland have continued to be regarded as reliable sources of information. As of 2019, among the 15 most opinion-forming media in Poland—five daily newspapers, three web portals, three TV news channels, three radio stations, and a free-to-air TV station (*Onet najczęściej*..., 2020-01-31)—only two were public.[11] The print media included three broadsheet daily newspapers (*Rzeczpospolita*, *Gazeta Wyborcza*, and *Dziennik Gazeta Prawna*) and two tabloids (*Super Express* and *Fakt*). The most often cited *Rzeczpospolita* received 29,100 quotations, followed by *Gazeta Wyborcza* (25,800), *Super Express* (15,600) and *Fakt* (14,500). An analogous ranking, analysing the impact of weeklies and biweeklies, revealed that in 2019 the top five of those were cited altogether only 14,100 times, that is, less than *Fakt* alone, with the leader receiving fewer than 4,000 quotes.

To give a possibly comprehensive picture of the presence of urban common good in Polish print media from 1989 to 2019, this analysis is based on the study of nine titles. *Rzeczpospolita* and *Gazeta Wyborcza* were selected as representatives of high-standard Polish dailies, while *Fakt* and *Super Express* cover the tabloid category. Despite their relatively lesser impact on public opinion, three weeklies were also taken into account. The choice of *Gość Niedzielny*, *Polityka* and *Newsweek Polska* relied on the criterion of the highest sales figures between 2005 and mid-2015 (Kurdupski 2015). Finally, quarterlies, that is, *Autoportret*, and *Res Publica Nowa* give an insight into the discourse on the margins of the mainstream.

*Rzeczpospolit*a is a liberal-conservative journal specialising in economic and legal matters, particularly targeted at highly educated professionals, managers, and high-rank public officers. Established as a press organ of the communist government in 1982, it became an independent outlet in the early 1990s. *Rzeczpospolita*'s main rival, *Gazeta Wyborcza* owes its name (Electoral Newspaper) to the parliamentary election of 1989, as it emerged during the electoral campaign to serve the Solidarity committee and promote candidates of the opposition. Its masthead motto: 'Nie ma wolności bez Solidarności' (There is no freedom without Solidarity), coined by Pope John Paul II during his second official visit to Poland in 1983, is a manifesto of commonality in itself.[12] The newspaper is generally considered centrist, with an occasional left-wing angle. Its readers may be generally characterised as middle class. Apart from covering international and domestic news, the paper features regional editions in major Polish urban agglomerations.

The daily tabloid *Fakt*, launched in 2003 by the Polish branch of Axel Springer, quickly achieved the highest circulation of all dailies. Maintaining a populist stance and targetting less affluent and less-educated readers, at the same time, it strives to represent higher-end tabloid journalism, complementing low-quality and sensational content with pieces by well-known experts and journalists (Filas and Płaneta 2009). On the contrary, *Super Express*, present on the Polish press market since 1991, mainly feeds on scandalous, provocative content, and does not refrain from topics considered politically incorrect or ethically dubious.

Gość Niedzielny is the only one among the nine titles reminiscent of the former counterpublic sphere; it is closely linked to the Catholic Church. It also has the longest history, dating back to 1923. Although its range is countrywide, each edition of the magazine includes local pages dedicated to news from twenty dioceses. The overall thematic focus is on issues of faith and religious overtones are present in most of the content. The *Gość* reaches audiences similar to *Fakt*'s, as the Catholic intelligentsia tend to read the more cultured and refined *Tygodnik Powszechny*. The remaining two weeklies differ mainly along ideological lines and areas of expertise. *Polityka* was established in 1957 as a ruling-party propaganda vehicle. In 1989, it began its independent life, ultimately navigating to its today's position, slightly left of the centre. *Newsweek Polska*, which first appeared in 2001, has fluctuated from conservative to liberal and centre-right to centrist. As much as *Polityka* takes much interest in social issues, *Newsweek Polska* is more into business affairs.

Autoportret, one of the two niche quarterlies selected for this analysis, has been published in Cracow since 2002. Subtitled 'a magazine about good space', it presents an interdisciplinary outlook on architecture and urban planning, viewing space as a cultural phenomenon. *Res Publica Nowa*, on the other hand, is a continuation of an, at first, clandestine, then legalised, *Res Publica*, issued from 1979 to 1992. Relaunched in 1992 under the 'renewed' name, it is a socio-cultural periodical with a more general programme than *Autoportret*. However, from 2009 until 2017, it featured a separate section devoted to cities, culture, and politics.

Since the period subject to this research spans three decades, when selecting the material for the analysis I had to rely mostly on digitalised resources of the nine outlets. Due to this limitation, the content published before the advent of the internet may be underrepresented, as some print media did not offer retrospective archives from the pre-digital era. Even in the case of the *Gazeta Wyborcza* repository, the only one claiming to contain all articles ever published in print, the less recent content comes up in searches only sparingly. Another factor which may have impeded the obtained results and their comparability is the forced variation of search techniques, dependent on the functional design of the different outlets' online services. Wherever these did not provide such an option at all, advanced Google searches were conducted instead.[13]

The search for articles published in the nine selected print media outlets relied on the combination of words 'dobro/a' and 'wspólne' ('good' and 'common(s)'), with results whittled down to hits relating to urban issues. The search outcomes yielded 239 articles, which were next coded with the use of the MAXQDA software in an inductive procedure of assigning categories to fragments of text containing both the searched term, and its context. The adopted grounded theory approach allowed me to categorise the data in a rigorous yet flexible manner, starting from open exploration and analysis of the gathered data to arrive at patterns extrapolated from individual cases by means of a relatively objective and unbiased analytic process (Thornberg and Charmaz 2013: 153).

The subsequent analysis partly relies on results of a pilot work undertaken in preparation for a more detailed study presented in this book, covering a shorter period of 2006–2015 (Grabkowska 2018). One of its key findings concerns a consistent expansion of urban common good in newspapers and magazines. As evidenced, in a majority of the six then-analysed titles,[14] the employment of the term 'common good' used in the urban context surged post-2011, the year which proved to be a watershed moment. The current research not only puts more print media titles under investigation, but the temporal scope of the analysis is extended as well. Although my original intention was to cover the whole transformation period, that is, the three decades from the beginning of 1990 to the end of 2019, due to the fact that a majority of the nine selected outlets have collected their published sources since around the beginning of the 2000s, it is only the last two decades that I could look into in more detail.[15] Those decades were further divided into four five-year subperiods to better trace the effect of the 'urban turn' around the 2010s.

In purely quantitative terms, obtained data revealed a growing trend in the incidence of the keywords within each timeframe (Figure 3.1). The overall number of articles mentioning common good in the urban context breaks down to 18 in the 2000–2004 subperiod, followed by 29 in 2005–2009, 94 in 2010–2014, and 98 in 2015–2019. In a majority of the newspapers and magazines, the numbers peaked or achieved the highest gains in the first half of the 2010s, while in four of the titles, the keywords had not appeared at all before then. Both findings from the pilot study—concerning the general upward trend, and locating the turning point around 2011—have thus proven valid for a twice-longer stretch of the transformation period.

A distribution of citations analysis in the print media sample should of course take into consideration their varying publication frequencies and problem-thematic horizons. Out of these two factors, the former explains why the use of the phrase '(urban) common good' was much more common—in terms of absolute numbers—in broadsheet dailies than in weeklies and quarterlies. The latter justifies the rift between high-quality dailies and tabloids, as well as the similarity of results for 'general' weeklies and 'specialised' quarterlies. Additional challenges to comparability follow from the already signalled differences in the applied search logarithms. Nonetheless, the juxtaposition of the gathered data is justified if we look at the general dynamics over the four subperiods.

1990–2004

Until around the first half of the 2000s, the searched keywords came up in rather general debates, usually contextualising common good as a guiding principle for local authorities and/or addressing the issue of communal resources. During this period, the narrative of the city as a commodity is born, with citizen participation being reduced to a bare minimum. In a telltale text penned under the title 'Self-governing—means free(d)', Jerzy Buzek, the then prime minister, distinguishes two basic types of local policymaking (2001-08-30). The first one is about a simple redistribution of state subventions. The second—endorsed by Buzek—attributes a much more operative role to the local governments as 'fully entitled managers of their own resources' and 'investors harnessing their own and private partners' resources in projects for the development of communes, cities, powiats, and regions'. What is emblematic here is an almost complete omission of citizens as key actors in decision-making. They are included in this vision in an inert, objectified position, when the author points to another, rather patronising, function of self-government—'initiator or partner of NGO initiatives serving the common good'.

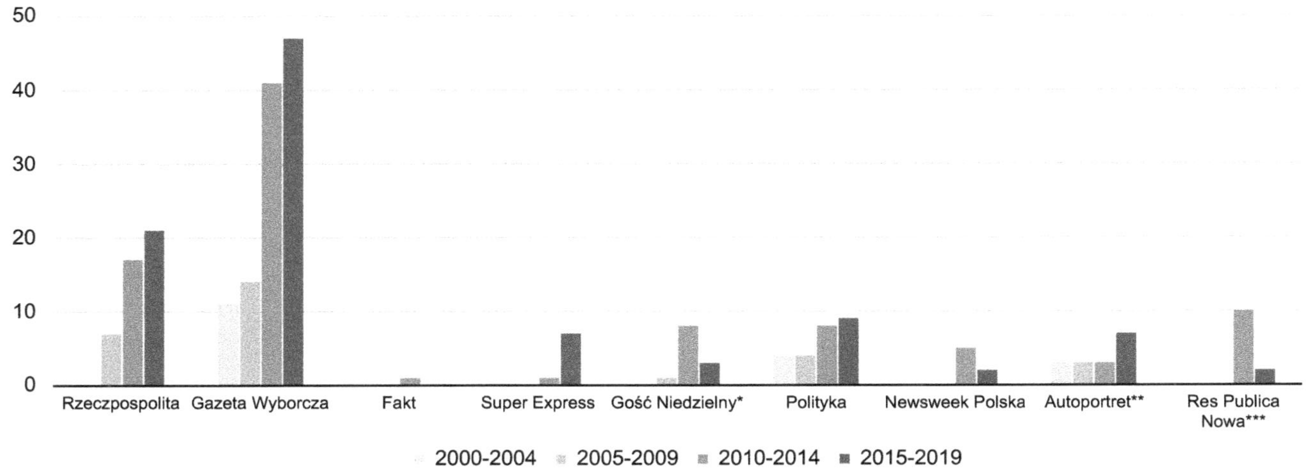

Figure 3.1 Number of articles featuring the term 'common good' in urban context within the analysed print media outlets between 2000 and 2019.

Whenever these 'early' articles touch upon the citizens' impact on the urban common good, it is usually in the negative context of deficient social capital. Urban dwellers are criticised for their non-compliance with laws and regulations concerning spatial planning (Szyperska 1999-01-16, Wielgo 1999-01-20), as well as for self-serving individualistic attitudes translating into spatial chaos, and a global ugliness of the urban landscape (Stefanowski 2004-08-12, Jaworski 2004-10-12). Another prominent topic relates to the disintegration of urban communities. Indicative here is an avalanche of readers' reactions to coverage of a conflict over a housing estate fenced off against the neighbouring municipal tenements in a medium-sized city. One of the quintessential comments to this article places the notion of common good in the service of anti-community sentiments: 'After all, anyone's blood would boil if they saw that their work was wasted and the common good was destroyed'.

Against this backdrop, an interview on the social effects of post-socialist transformation, reporting research findings which document a slow change in Poles' attitudes in favour of the local common good (Bogucka 1996-05-10), stands out as an exception that proves the rule. Altogether, the whole period is marked by mostly sweeping references to urban common good, with no systemic interpretations whatsoever. The city is dubbed common good for the first time in 2003—in the announcement of the Congress of Polish Urban Planning in Gdańsk (2003-07-30). The outcomes of the event, held under the theme 'City—common good and collective responsibility', were presented later just as briefly, and summed up with overly general remarks (2003-09-16).

2005–2009

Further into the first decade of the third millennium, the critical tone comes to the fore. Two distinct threads dominate the reflection on urban common good—its framing as 'no one's good' and as opposition to the 'sacred right to property'. They both add up to a narrative lamenting the neglect of urban space and infrastructure as a 'sad legacy of the Polish People's Republic' with the rise of privatism as a 'form of acting out the limitations of the former regime' (Wrabec and Socha 2006-03-25). An analogous metaphor of 'decompression' after the long decades of virtually non-existent property relations is used by two environmental psychologists explaining the excessive popularity of gated residential developments in Poland (Surmiak-Domańska 2007-10-01). Poles are found to be lured by developers' promises of enclosed Arcadias, or sanctuaries, where they can cultivate their private spaces oblivious to public misery and disorder outside the guarded fences. The unleashed processes of residential segregation are also interpreted as a revanchist response to the formerly enforced social mix.

Some authors, seeking to uncover other factors responsible for an elevated status of property rights after 1989, link this rationale to the abandonment of local planning with the advent of free market ('There was one single rule: build wherever you like and however you like'), pairing it with the discredit of the communal ideals ('Whoever uses land, and becomes its owner, can do as they please with their sacred property, and whoever thinks otherwise is a proponent of communism') (Bartoszewicz and Jarecka 2008-07-19). Some other ones find it to be justified by the ideology of individualism, promoting the idea of equal opportunities in working towards one's own accomplishment and prosperity. According to one of the best-selling Polish reporters, personal success 'is defined through property' which not only becomes a 'simple, if not simplistic tool of discriminating between the achievers and the underdogs' but also allows evading all responsibility (Springer 2016-06-11). To others still, the sanctification of property rights is attributed to the ineffectiveness of the state, as well as to corruption and/or discretionary power of public officers. In this perspective, loose regulations are found to be serving the interests of property owners instead of protecting the ill-defined common good (Żakowski 2010-06-05).

Five years after the Congress of Urban Planning took place in Gdańsk, the notion of the city as a commons reemerged in the analysed print media—which shows how slowly the idea was breaking through. Recalling a meeting of the ministerial commission of urban planning and architecture, a Warsaw-based architect recounts the gathered experts' reaction when the minister of construction had suggested the superiority of the free market principle over urban planning standards:

> we explained that the free market does not take common good into account, and that the common good is the city. It cannot exist without the free market, but the main task of local government lies in mitigating conflict between the private and the public to achieve maximum consensus for the common good.
>
> (Bartoszewicz and Jarecka 2008-07-19)

In an essay on common space appropriation, published the next year in *Autoportet*, the same architect continues this train of thought. She first observes that a former 'theoretical dictatorship of the proletariat' is currently being replaced by a 'genuine dictatorship of the so-called free market and the mighty power of capital, of which the developer is a personification' (Staniszkis 2009: 82), which brings her to the conclusion that the indispensable condition for creating liveable cities is a combination of political will, and appropriate regulations, which act against voluntarism and in favour of the common good (83). This statement comes up as a harbinger of the city as a commons narrative, representing a standpoint which had already been making its way in the first decade of the 2000s, to ripen in the second one.

2010–2014

Undercurrents eroding the prevailing discourse of the city as a commodity were noticeable ahead of the watershed moment of 2011. In 2010, two urban activists from Poznań identified five such counterdiscourses in their local urban milieu (Mergler and Pobłocki 2010). The first undermines the logic of commodification of urban space and sacredness of the right to property. The second exposes the crony-capitalist lack of transparency in relations between power and business elites. The third criticises the immoderate scale of urban resources privatisation, comparing it to a 'sell-off of prized family silverware', and points to overall mismanagement at the local level. The fourth addresses the paternalism of local authorities, prompting them to interfere in all urban affairs from a position of superiority, and being the ones accountable for the superficiality of public consultations. Lastly, the fifth revolves around the neglected issue of sustainable development, underrepresented in the official urban discourse. All five counterdiscourses are found to be not only interlinked but also establishing 'a completely new semantic space, new notion of the common good and new vision of urban development' (9). Altogether they are expected to give substance to the concept of the right to the city, which according to the authors, had not, 'as of yet' been fully recognised by the nationwide media.

At around the same time, the print media switched attention to citizens, eventually recognising them as, at least potential, agents of urban change. This category is mostly represented by the middle class, albeit in multiple variations. One is that of 'new metropolitans' (*nowi wielkomiejscy*)—newcomers to big cities in pursuit of careers, with self-centred attitudes and low citizen awareness proving an impediment to the appreciation of urban common good (Lichocka 2010-07-20).[16] A complex portrait of middle-class citizens from the other end of the spectrum—self-organised and mobilised to take collective action—is provided by Lech Mergler (2012). Being a local activist himself, he convincingly depicts them as 'simply inhabitants' of Poznań prompted to engage in 'war over urban space' whenever they find the wellbeing of urban dwellers and city users threatened by local power coalitions of business elites and authorities (45). A similar group—conceptualised by Paweł Kubicki, a sociologist—under the name of 'new bourgeoisie' is ascribed noble intentions of seeing the city 'not as a sum of private properties but as common good' (Klich 2011-06-04).

Considering the middle-class bias of the media sample, the omission of the perspective of the lower social stratum in these considerations is hardly astonishing. More difficult to explain is the fact that urban movements, which shook the urban arenas in Poland at exactly that moment, tend to appear somewhat in the background of the debate on urban common good. Only a few texts within the analysed sample relate to their role in restoring the latter (Pobłocki 2015-05-20):

> [owing to Polish urban movements] people have noticed that in addition to being mothers, fathers, Poles or Catholics, they are also inhabitants, and that common good is not an abstract idea nor that public life is a 'celebration of democracy' which happens only every four years, but rather something very specific—a park, a school, or an empty square outside the window.

A relative absence of urban movements contrasts with an overflow of articles addressing the already mentioned issue of low social capital in urban settings. One of the recurrent topics is the anti-social behaviour of urban dwellers—ranging from not cleaning up after dogs to disengagement from local community (Giedrys 2014-03-29). It is well-depicted by a *Newsweek Polska* columnist, who observes, in a piece with a hard-hitting title of 'Polish Poo', that—unlike in western Europe—in Poland, it is 'unsafe' to lie down on urban park lawns. He explains this by the dialectics of 'no one's good' versus 'sacred right to property':

> [In Poland], some people on leaving their flats feel as if they step into a social black hole, where they have no rights, duties, or responsibilities—as if on their way from home to work, to the cinema, they crossed 'international waters'. Homes, companies, shops must look nice. But what is common, to many still means nobody's.
>
> (Hołownia 2011-03-12)

The comparison to western European cities is emblematic here. Several years into Poland's membership in the EU, the author expects his compatriots to live up to the 'European' standards in terms of care of the urban common(s). Similarly, critical diagnoses concern the Polish 'tradition' of collective action for common good only happening in bursts (Nizinkiewicz 2013-08-30), the logic of solidarity being currently replaced by a dog-eat-dog mentality and 'consigned to the dustbin of history' (Makowski 2013-09-16), and Poles' deep-rooted individualism as a reminiscence of the low esteem of common good in the times of the noble republic (Pacholski 2014-03-08, Matys 2015-03-14).

In spite of this rather pessimistic outlook, there are also signs of a shift. First of all, the articles comprehend and promote the importance of citizen participation. Participatory budgeting comes to the spotlight, receiving high praise in all outlets. A short article in *Gość Niedzielny* reports its implementation in one of the middle-sized Lower Silesian cities, citing the mayor's enthusiastic account of an 'impressive' number and quality of filed applications, which he regards as clear indicators of the inhabitants' needs and perceptions (2014-01-02). In *Gazeta Wyborcza*, an interviewed political think tank leader linked to the Civic Platform underlines the educational role of the tool:

> The amount [of allocated funds] does not really matter that much (…). The point is for people to understand that the city is their common space and common good.
>
> (Lipoński 2013-11-08)

On the pages of *Rzeczpospolita*, the same expert calls for a 'reactivation of the culture of common goods' through a return to the principle of social solidarity (Makowski 2013-09-16). He proclaims the latter to be an antidote to social inequality and an overall effective alternative to extreme individualism. A similar but more concretised and inclusionary appeal is voiced by an architect and urban planner in an essay penned for the 2015 Civic Congress, reprinted in *Rzeczpospolita*:

> We need [...] to take into account the needs of different city users (including those with low credit rating). [...] And [in the city] we need more than accommodation, workplaces and roads linking them. We need warranties for common goods in urban space and we need conditions for predictability. Therefore, we need regulations which will enshrine them.
>
> (Kipta 2015-06-01)

Another prominent theme built around the central idea of urban common good is support for collective action for common good rising above political divisions. An article in *Gość Niedzielny* quotes a conservative historian maintaining that political polarisation should not preclude active citizenship, as the latter 'has no political colour' and 'works great in small, but no less important matters' of urban common good at the scale of the city or neighbourhood (Legutko 2013-08-29). A similar opinion is expressed by a non-partisan mayor of Gdynia, who recommends discarding political affiliations at the local level and 'treating the urban organism as a common good' (Szczurek 2011-03-03).

Finally, a considerable portion of the analysed texts dating from the 2010–2014 period is devoted to an outpour of various manifestations of urban commons. Commoning comes up in articles on uprising trends of cohousing (Czyńska 2010-11-13, Lipczak 2014-04-12) and flat-sharing (Szwed 2012-06-23), discussing the future of allotments (Wielgo 2013-02-26), and emerging food cooperatives (Szymanik 2013-07-25), as well as in a piece on a pay-as-you-wish community café in Kraków (Hawranek 2014-01-16). While the tone in most of these texts is informative and well-disposed, a few employ irony and sarcasm. For instance, the story of the food co-op abounds in mockery, deriding the collectivist rules over obligations, contributions, and decision-making adopted by the founders, and terming the battle against the slugs which feed on the crops 'combating the counter-revolutionaries' (Szymanik 2013-07-25). Its title—'A carrot in utopia'—not only conveys the author's lack of faith in the success of the enterprise but wholesale ridicules the idea.

Apart from simply describing commoning practices, some articles place them in the context of the fight for common good. A feature on an initiative of pensioners who form a community vigilante group to take care of the then run-down Wojewódzki Park Kultury i Wypoczynku—tracking vandals and planting trees in community service (Malinowska 2010-02-17)—provides one such example. Another relates to the findings of a research report on the functioning of microcinemas in Poland—being recognised in small towns as common goods, any threat to their further existence mobilises local communities to come together in a joint struggle to ensure their survival (Popławska 2013-05-24).

2015–2019

By 2015, urban common good no longer appears as a utopian or novelty idea in the analysed print media. While it remains on the margins of the mainstream, its position is well-established—the idea is invoked by leftist politicians (Wodecka 2017-04-29) and conservative urban planners (Bielecki 2015-09-05) alike. According to one publicist, the rediscovery of urban commonality springs from a rising awareness of 'interdependence, the need to rely on others' on the one hand, and an increasing sense of agency on the other (Maciejewski 2017-03-11). While he considers the former to be 'the essence of urbanity, mak[ing] us experience community as something real and tangible', the latter allows for feeling the impact on the world around us in the most direct way—'e.g., by blocking removal of a tree, enforcing construction of a pavement or cycle lanes'.

The articles from that time mostly credit urban movements but also associate the turn towards the urban common(s) with Poland's accession to the EU. The urban planner Ewa Kipta (2015-06-01) argues that post-accession migrations, which had enabled the Poles to fully experience urban living in Western Europe, consequently launched 'a process of rebuilding the collective understanding of urbanity' in Poland. She considers this process to be happening intrinsically, as a result of the growing appreciation of urban commonality values:

> Today we are maturing to the fact that life quality is shaped not only by private spaces, in which we spend less and less time. We have found ourselves in need of good common spaces to connect with other people. Because the essence of cities is proximity and ease of [social] interactions [...]. We are slowly noticing the shortcomings of excessive expansion of private spaces and too strict segregation of urban functions, which forces us to commute daily. Increasingly, we begin to appreciate the complex, 'inconsistent' structure of older inner-city neighbourhoods, the urban markets, streets, and squares. We take to cycling and make use of the green space.

Whereas some urban dwellers may be excluded from this vision, similar optimism bubbles up from an assessment by an activist from Warsaw, who anticipates that 'the alliance of city activists with lemmings [in a joint fight for the urban common good] is only a matter of time' (Śpiewak 2015-04-18). Pointing out their common agenda—criticism of public policies at the central and local level, and recognition of the relevance of urban common good—he deems reconciliation of the two groups conceivable despite their seemingly conflicting programmes. He also indicates improvement of the quality of public space as the potential 'first common goal' and, therefore, a possible catalyst for cooperation.

Importantly, the common good tends to lose its idealistic quality. Instead of being a given once and for all 'constant primordial', it is increasingly perceived as a process of constant renegotiation of 'an ever-changing set of interests of all the people operating in the city' (Gzowska 2019), coming with much responsibility and requirement for regular engagement (Mazur 2015-07-23). A related trend consists in a ripening debate on tools of governance. Reflection on participatory budgeting acquires more insight. Its novelty wears off, exposing flaws in the formula adopted in Poland—lack of deliberation, shifting responsibilities of the local government to citizens, delays or failures in the implementation of the winning projects, copy-paste ideas, and so on (Kmieciak 2018-01-25). Some articles deliberate possible ways to upgrade the method, while others look for alternative solutions. One of the latter investigates the potential of civic assemblies, quoting the participants of its pioneer editions and experts, who both recommend this approach as decision-making 'by people not aiming for the next elections but for common good' (Socha 2017-05-30).

Yet, aside from multiple signs of the city as a commons paradigm settling in, the right to the city remains contested. One group of opponents are experts rejecting it on the grounds that uninformed citizen participation aggravates spatial disorder (Springer 2016-06-11). Another consists mostly of unwavering followers of neoliberal philosophy, and the ideology of privatism. Their arguments are either based on the claim that the private sector and individual initiative are more effective than public policies and collective action (Majcherek 2011-08-22), or simply consist in clinging onto the sacred right to property. In the article 'Will Taxpayers Finance Trees Planted on Private Plots?', the author ridicules 'the absurdity' of protesting 'lex Szyszko':

> If a tree from a private property is a common good, then we should all contribute to it in taxes. Because why not? [...] The question is: will those who are protesting today be equally willing to pay from their taxes for the costs of planting, care and time spent by the owner on necessary gardening?
>
> (Teister 2017-02-23)

Proponents of the idea of common good explain such disregard for urban common(s) as a long-term consequence of the 'rejection of the heritage of Solidarity' (Maciejewski 2017-03-11) and indiscriminate adoption of 'the neoliberal ideology, fatal for the quality of social life' (Szahaj 2017-02-09). They also point to the problem of citizens giving up on political engagement in local affairs, and passing the buck on responsibility (Gzowska 2019).

Apart from the general reflections on issues related to the urban common good, a considerable portion of specific urban 'common goods'. The results of the 2006-2015 pilot study point to a versatility of this substantive interpretation of the term common good, employed to designate an assorted range of urban resources—from façades and backyards, through streets, parks and squares, to urban landscape and spatial order (Grabkowska 2018). For the purposes of this research, I arranged different understandings of urban

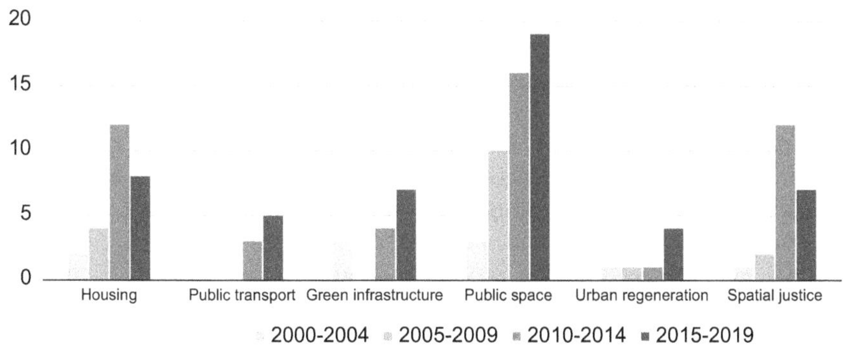

Figure 3.2 Number of articles featuring the term 'common good' in six thematic areas within the analysed print media outlets between 2000 and 2019.

common goods in the six thematic categories discussed in Chapter 2. In total, they were mentioned in 126 articles, displaying a general upward trend in terms of occurrence (Figure 3.2). A simple quantitative analysis reveals that over one-third of them (38.0 per cent) related to public space, around one-fifth to housing (21.4 per cent) and urban justice (17.5 per cent), followed by green infrastructure (11.9 per cent), public transport (6.3 per cent), and urban regeneration (4.8 per cent).

The most numerously represented category of **public space** features an overly critical perspective. While such tone pervades the texts throughout the whole analysed period, it is particularly pronounced during the first two decades. Public space in Polish cities is referred to as 'ruled by money instead of law' (2003-09-16), terrorised by graffiti artists (Stefanowski 2004-08-12), treated as irrelevant (Jaworski 2004-10-12), shrinking (Wrabec and Socha 2006-03-25), enclosed (Podgórska 2007-01-06), unclaimed (Bielecki 2012-01-14), unfriendly (Wybieralski 2013b-11-29), appropriated (Szahaj 2017-02-09), and littered with visual advertising (Szahaj 2017-02-09). The narrative on post-socialist transformations of urban space builds on general remarks on the spatial disorder in cities, as well as on privatisation and commercialisation of public space. While the still recent socialist past is mostly held responsible for the former, the unfolding capitalist present tends to be viewed as the root of all problems connected to the latter (Wrabec and Socha 2006-03-25).

Since around the mid-2010s, the perception of the city and public space as commodities begins to bend to the communal way of thinking. Some publicists attribute this to the realisation that the quality of urban space is an economic value and that 'it is worth to protect [this] common good, because it pays off for all of us, also individually' (Krupa-Dąbrowska 2013a-06-10). But there are also testimonies of alternative rationales taking over: the ruling of the Supreme Court that an unauthorised reduction of window openings undertaken by a tenant in a multiple-occupancy building is unlawful, on the grounds that the façade is a common good (Krupa-Dąbrowska

2013b-11-22), or a massive public outcry against the phenomenon of 'beach windscreening' (Koźlenko 2014-08-19, Maciejewski 2017-03-11.

Articles concerning the issues of **housing** present it as equally threatened by social disinterest and lack of concern (the narrative of 'no-one's good'), and appropriation (the narrative 'the sacred right to property'), with little hope for evading either. Both perspectives dominate throughout the whole period, illustrated by two contrasting images—the devastation of a newly built council housing unit in Warsaw, and enclosures of urban space through residential gating. Similarly grim is the undertone of analysis of the role of foreign investors as stakeholders in urban development. According to the architect Magdalena Staniszkis (2009: 82), who draws a line between 'co-building the city' and 'treating the city as a construction site', the difference lies in the way developers relate to the common good of local citizens—whether they choose to respect it or treat it instrumentally. Another text conveys the message that common good fails to be safeguarded even in such established communal institutions as housing cooperatives, which are presented in a negative light, their workings non-transparent and potentially corrupt (2011-11-24).

Housing issues resonate in many a text that fall into the broad category of **spatial justice**. One of the articles, bringing up the problem of forced tenant eviction practices, contrasts the 'resourcefulness' of private landlords, effectively defending their individual interests, with the ineptitude of tenants trying to exercise their right to housing (Pobłocki 2012-10-19). In a similar vein, an activist interviewed in *Gazeta Wyborcza* recalls how public internet fora users tend to refer to evicted tenants as 'idlers, alcoholics, bums who impede management of the sacred private property' (Wybieralski 2013a-01-19). Around the mid-2010s, the ills of reprivatisation processes ongoing in Poland are explicitly termed a 'long-running outrage' (Springer 2016-06-11). The author juxtaposes them with the anti-gentrification measures implemented at the time by the local authorities in Berlin.

Other articles discussing contemporary problems of Polish cities refer to the idea of justice in the context of local governance quality and inclusiveness (Jarkowiec 2016-10-15, Gzowska 2019), discrimination in the access to different uses and effects of transformations of urban space (Żakowski 2010-06-05, Milewska and Pacewicz 2018-09-21), social inequalities related to air pollution (Mucha 2017-10-12). However, what are missing from the debate spanning the whole analysed period is the non-middle-class perspective and the issue of the right to the city of the marginalised socio-cultural minorities and immigrants.

In the analysed period, the debate on the communal values of **green infrastructure** initially revolved mostly around allotment gardens, presented as relics of the communist past facing the pressures of the unfolding neoliberal agenda. Since around 2010, considerable attention has been given to informal initiatives for urban common greens. Articles feature stories on specific collective actions undertaken by citizens, such as the aforementioned vigilante patrolling of the Silesian Park (Malinowska 2010-02-17) or micro-scale guerrilla gardening in Kraków (Adamczyk 2015-11-12). Interestingly, as much as the bottom-up engagement is applauded by the correspondents, it encounters resistance from local officials, whose reception is reported to be far from enthusiastic.

An extended debate on the communal dimension of green infrastructure and the clash of private and collective interests can be found in articles discussing 'lex Szyszko'. Apart from the already quoted defence of the right to unlimited control over 'private trees' (Teister 2017-02-23), opinions gravitate towards acknowledging the need for safeguarding common good. The title of one of the articles calling for a systemic legal protection of trees—'The Polish Chain Saw Massacre' (Sapała 2017-02-28)—paraphrases the title of a well-known Hollywood horror film series.

Earlier articles concerning **public transport** point to globally changing standards—favouring collective modes of moving around in cities—and discuss the reasons why such standards do not seem to catch on in Poland. One of the texts begins with an observation that 'in Polish cities, it is not transport that adapts to the existing urban layout, but the urban fabric conforms to the needs of drivers' (Beim 2011-05-07). Another diagnosis, specified in the name of an invented disease in the eponymously titled 'Carosis Polonica' (Pacholski 2014-03-08), concentrates on the reasons behind the unfaltering popularity of individual car transport in Poland. Part of the blame is laid with local governments for 'taking the easy way out' and adopting the let-citizens-take-care-of-themselves principle instead of providing more efficient public transportation systems.

Accounts signalling a change of paradigm in the domain of urban transport first emerge around the mid-2010s. An expert asked about possible remedies to the problem of traffic congestion in Poland mentions not only innovative solutions from Western Europe but also best practices from the socialist era, especially in terms of spatial planning prioritising pedestrians and public transport (Kompowski 2016-10-21). The article's title—'City for the People, Not Just for Cars'—draws from the slogan of a social campaign organised by *Gazeta Wyborcza*. Towards the end of the analysed period, these repeatedly discussed topics give way to more recent challenges. One of them consists in the absence of culture of common good impacting the emerging sharing economy. For instance, an article in *Super Express* encourages readers to contribute their opinions on problems generated by the newly introduced scooter-sharing schemes in Poznań (Brzeziński 2018-12-19).

Urban regeneration takes up the least space in the debate on urban common good, which partly results from the fact that revitalisation processes gained momentum only after Poland's accession to the EU. Most of the few articles on the topic included in the sample either criticise insufficient social impact of urban regeneration programmes (Hac 2014-08-28) or pay tribute to architectural and communal values of completed infrastructural projects (Sarzyński 2015-05-12). More recently, the trending narrative of urban commons has been hijacked by developers. In a sponsored article of 2018, a Warsaw-based property development company investing in modernisation of old tenement buildings describes its mission as 'acting as an educator' by making the public aware of the 'beauty and potential' of the 'architectural pearls [which] have been falling into oblivion' (2018-06-04).

Even though the analysed sample offers only a glimpse into the print media discourse of the last three decades, it is evident that its impact was much greater than in the case of the investigated body of legal texts. A chronological overview of the articles documents the initial disintegration of the post-socialist urban common, when individualism, privatism, and the sacred property rights took over, and the ensuing fight for the right to urban common good, when the paradigm of the city as a commodity rivalled that of the city as a commons. While the overtone of the early pieces is mainly negative and the emphasis is put on the deficiencies of the Polish cities in transition, the more recent ones tend to approach the question of urban common good from a more constructive angle. The levels of mobilisation of citizen engagement and collective action appear to vary across the six thematic categories examined in more detail. However, the 'reactivation of the culture of common goods', postulated by Jarosław Makowski (2013-09-16), seems to have already taken hold in Poland.

Urban common(s) as embraced in academic research

In Poland, recent years have brought a host of monographs on the overarching theme of urban commonality. The most notable example is Paweł Kubicki's (2016) book on the current reinvention of urbanity in Poland, recounting the changing position of the city in the system of values and public discourse over several centuries of the country's troubled history. A considerable number of other latest scholarly works relate to processes of urban transformations ongoing in Polish cities within the six thematic areas identified and discussed in Chapter 2. They include the commodification of housing and reclamation of the right to affordable accommodation (Leśniak-Rychlak 2018/2019, Twardoch 2019, Erbel 2020), shrinking public transport networks (Gitkiewicz 2019, Trammer 2019), vanishing green infrastructure (Mencwel 2020), transformations of public space (Dymnicka 2013, Bierwiaczonek 2016), as well as the broadly defined problems related to spatial justice (Bukowiecki et al. 2014, Jacobsson and Korolczuk 2017, Sagan 2017, Domaradzka 2021). Another group comprises Polish editions of the already 'classic' works by foreign authors, such as Jan Gehl (2009), Michael Hardt and Antonio Negri (2012), David Harvey (2012), Charles Montgomery (2015), Jane Jacobs (2017), or Janette Sadik-Khan and Seth Salomonow (2017).

While the list is not exhaustive, such proliferation already indicates a significant interest of the academic and expert milieus in matters revolving around the broadly defined urban common good. As for more direct references to this notion, it was a text penned by Łukasz Pancewicz (2013) which inspired me to take up a research project that this book sums up. Published in a post-conference volume compiling the contributions of participants in the 6th Pomeranian Civic Congress, it accounts for a concise, yet comprehensive, analysis of the city as a commons.[17] Drawing from the works of Lefebvre, Harvey, and Ostrom, the author—an architect and urban planner—identifies commoning as a new model of urban resource management, counteracting fragmentation and commodification, as well as improving urban life quality, and ensuring spatial justice. The publication concludes with an appeal for not confining the issue of common good to 'the sphere of academic debate, moral imperative or "symbolic politics"', as it would reinforce the abstractness of the term, and render it a worn cliché, useless, and unproductive (Pancewicz 2013: 17).

The following analysis however aims to focus on this very academic debate to observe how it absorbed the notion of urban common good and the paradigm of city as a commons, as well as to assess the capacity to transmit them further into the public sphere. Four main selection criteria were applied to that end. Firstly, the publications had to have a potentially high impact on public discourse. Secondly, they had to employ the term common good and/or commons in the context of Polish cities after 1989. Next, they had to be published between 1990 and 2019. Lastly, they were to be written in Polish or English—as the latter has grown to become the academic lingua franca in Poland.

Measuring the actual impact of scientific research on public discourse is a problematic task. To approximate 'influentiality' of the texts to be included in the sample, I relied on their prevalence in online academic search engines—widely accessible in terms of recognition and being free of charge. To maximise reach and relevance, four different online bibliographic tools were used: Biblioteka Nauki (BN, http://yadda.icm.edu.pl), Central and Eastern European Online Library (CEEOL, http://www.ceeol.com), Google Scholar (GS, http://scholar.google.com), and Polska Bibliografia Naukowa (PBN, http://pbn.nauka.gov.pl). Among the four, GS is the most popular and extensive. However, as a crawler-based web search engine and not a typical bibliographic database, its accuracy and reproducibility are questionable, and it is therefore recommended as a supplementary source of evidence (Gusenbauer and Haddaway 2020). CEEOL is a renowned e-repository of academic books and journals specialising in the humanities and social sciences native and/or dedicated to Central and Eastern Europe. The other two databases are run by Polish academic and governmental institutions—BN by the Data Science Centre at the University of Warsaw and PBN by the Ministry of Education and Science. They were used to ensure sufficient representation of local scholars' publications.

In April 2021 the four databases were searched for several keywords, namely: 'city' AND 'common good'; 'city' AND 'commons'; along with their Polish equivalents—'miasto' AND 'dobro/a wspólne'.[18] Whenever the initial search case returned more than a hundred of items, they were narrowed down by the addition of 'post-socialist' or 'postsocjalistyczny' and 'Poland' to the search terms. If the resulting number remained high, only the first hundred would be taken into consideration. Where possible, the search options were set to include not only journal articles, but books and book chapters as well.

Project reports were eligible, and unpublished papers were not. Each of the search results was thoroughly examined in terms of the other selection criteria. A few publications were eliminated at this point since their content did not match the context of this study.

A considerable number of results obtained from different databases overlapped, which indicates that the decision to employ several search engines worked in favour of sampling exhaustiveness. Altogether, the compiled list of results comprises 36 publications, that is, 22 journal articles, 9 book chapters, 4 books, and 1 project report. Their authors are mostly representatives of social sciences—urban geographers, sociologists, economists, and legal scholars decisively outnumber architects and urban planners. Twenty-three works, almost two-thirds of the total number, have been penned in Polish. Although the earliest publication dates back to 2008, over a quarter of the texts were published in 2017 (Figure 3.3). The upward trend mirrors the one observed in the analysis of the print media discourse, but with a clear time lag of a few years. This finding confirms the assumption that changes in the print media discourse precede changes in the academic discourse.

The selected sample mainly entails literature reviews and publications of theoretical character. An array of research topics is covered, from broader categories of civic society and civic participation (Domaradzka and Wijkström 2016, Popow 2016, Sobol 2017), social justice (Bugno-Janik and Janik 2019), and sharing economy (Polko 2017), through problems of spatial order (Szewczyk 2018), urban landscape (Wiszniowski 2019), and housing cooperatives (Matysek-Imielińska 2018), to such specific issues as the real-utopian potential of urban commons (Bugno-Janik et al. 2019). A few of the more recent works adopt empirical approaches, exploring specific problems through quantitative or qualitative research. Here, the investigated issues include Polish citizens' attitudes toward the common good (Borowik 2015, Sobol 2016a, Bednarska-Olejniczak and Olejniczak 2018), as well as public discourses related to social cohesion and diversity (Korcelli-Olejniczak et al. 2017), and to urban space viewed as a commons (Grabkowska 2018).

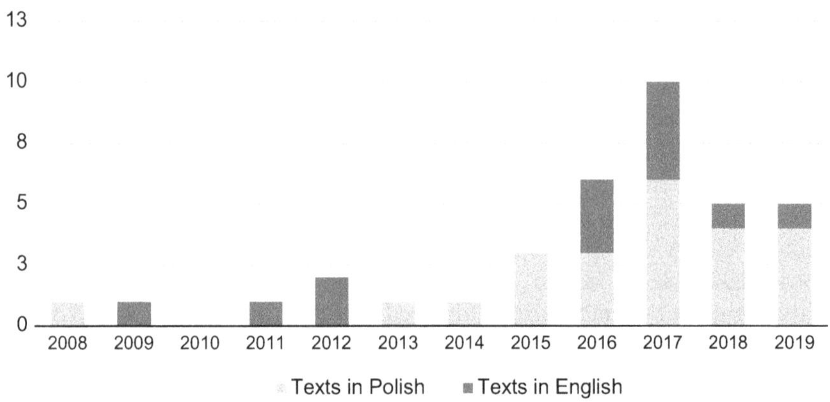

Figure 3.3 Analysed academic works by year of publication.

This transition from theoretical considerations towards more in-depth studies is also indicative of a progressing concretisation of the analysed concepts over the research period—from 'rediscovering the common good' (Słodowa-Hełpa 2015), through identification of 'creators of urban common goods' (Czornik 2017), to such pragmatic deliberations as examining how urban commons utopias could be turned into reality (Bugno-Janik et al. 2019). Another tendency is the apparent thematic split within the selected works—some publications focus on (urban) common good while others foremostly deal with the paradigm of the city as a commons. I will first look more closely at the first group and then move on to an investigation of the latter, although both perspectives often intermingle.

Polish cities and the (urban) common good

Most of the authors referring to common good in the urban context are much aware of the term's elusiveness, and they underline its 'generality' (Smarż 2016: 254) or 'broadness' (Sosnowski 2017). To Słodowa-Hełpa (2015: 8), this capacity, 'reflecting the diverse social contexts in which it arises and functions', accounts for power and universality as much as for limited comprehensibility. In turn, Bugno-Janik et al. (2019: 35) maintain that any strict and once-and-for-all definition is impossible and even undesired for many reasons—a heavy dependence on local conditions being just one of them. Quite many academics who attempt to situate the urban common good within a theoretical framework go back to the works of Greek philosophers, Aristotle in particular (Sobol 2016b, Grabkowska 2017, Sobol 2017). Others prefer to cite catholic scholars (Słodowa-Hełpa 2015) or integrate both schools of thought (Machocka and Śwital 2016)—it needs to be acknowledged that this reliance on the social teaching of the Catholic Church, quite typical of Poland, bears witness to a specific continuation of the alternative public sphere from the communist period (see Chapter 2). Numerous works include direct references to the Constitution and other acts of Polish law (Machocka and Śwital 2016, Smarż 2016, Sobol 2016b, 2017). The concept of concrete narrative elaborated by Mergler et al. (2013) represents the sole 'local' theoretical framework (Domaradzka and Wijkström 2016, Popow 2016).

The earlier publications tend to employ the concept of urban common good in a more general manner. In a short paper on the urban design of new housing estates in Poland Maciej Kowaluk (2011) juxtaposes common good with maximisation of developers' interests, without any definition whatsoever. No explanation of the term is provided, nor any further elaboration appears in the text, apart from a vague reference to public good, which is apparently treated as a synonym. Likewise unspecified 'common good' serves as a litmus test for measuring civic involvement at the local level in the surveys conducted by Zinserling and Grzelak (2008) and Kotus et al. (2019). Both studies underline the need for assessment of citizens' willingness to engage for the sake of the immediate surroundings and their disposition to intervene whenever (undefined) common good is at stake.

Several of the more recent texts combine the idea of urban common good with urban commons. For instance, Domaradzka and Wijkström (2016: 303) find evidence for urban activists' perception of common urban space both as a common-pool resource *and* a shared value. Correspondingly, Słodowa-Hełpa (2015: 16) proposes a three-fold definition of the common good—very much in line with the model adopted in this book:

> The paradigm of common good connects communities with a range of social activities, values, and norms used to manage common resources. In other words, it comprises three elements that make up an integrated, interdependent entirety: resources, community, and a set of the above-mentioned principles, values and norms defined by social protocols. The general principles include participatory democracy, transparency, honesty, personal access to benefits, manifested in practice in very individual and specific forms following from a reflection on what is to be achieved, what is the social value of doing it.

One of the prominent threads in the analysed works concerns stakeholders engaged in the production and/or consumption of urban commonality. Opinions vary on whom to identify as leading actors and who plays supporting roles on the urban scene. For example, Machocka and Śwital (2016: 140) approve of city councillors' monopoly over decision-making, justifying it by the rule of law. According to the authors, citizen influence limited to participation in local referenda, local elections, and public consultations is sufficient. An opposing view is held by Borowik (2015: 99) who denounces local decision-makers as 'arbitrary' and 'hiding behind public and community interests', while Wiszniowski (2019: 20) disapproves of their 'disproportionately privileged position'.

Nawratek (2012: 96), on the other hand, blames architects and urban planners for considering 'themselves as depositaries of the universal "common good"'. He subsequently refutes this way of thinking on the basis of its exclusionary logic:

> [It is commonly believed that] common good (…) can only be recognised by appropriately prepared people—either through education (…) or because of the high position they hold, allowing them to see the world from an almost divine perspective. It is the moment of arrogance which in fact reduces the importance of democracy and 'people power', assuming that decisions should only be made by these who claim to understand the city. This is where democracy collapses when confronted with the common good.
>
> (Nawratek 2012: 99)

The crisis of representative democracy at the local level in Poland is a topic frequently brought up in this context. Multiple authors claim that urban regimes rely on technocratic and bureaucratic arrangements hampering civic participation and engagement (Dymnicka 2013, Sobol 2016b). Moreover, the responsibility for common good is found to be either shrugged off or shamelessly handed over to business actors—Borowik (2015: 99) uses the expression 'passing the buck' (*umycie rąk*) in this context. Overall, the observations in most of the accounts boil down to a conclusion that despite being 'guaranteed' in the Polish Constitution, co-governance of the urban common good is rather an impossible project.

The issues presented thus far bring us to an inevitable debate on conflicts of interests. A pro-liberal argumentation is provided by Smarż (2016), who claims that individual rights and freedoms should only be limited for the sake of 'the most important values, such as: life, health, natural environment and public safety' (256). Individual interest prioritisation over common good is addressed by Staniszkis (2012: 102), who sets forth a bitter diagnosis:

> If architecture reflects society, the Warsaw cityscape says a lot about the condition of society and the mental attitudes of architects in post-communist times. The contemporary syndrome of the culture of common good being trumped by the culture of individualism is plain in Warsaw.

In a similar vein, Sobol (2016b: 2) observes that 'a strong individualistic drive' in contemporary Poland is accompanied by 'a deficit of community initiatives and concern for common good in the everyday life of city dwellers'. The author associates this cultural shift with the historical context, claiming that the fact that the declared concern for the common under socialism in practice came at the cost of civil liberties, which explains today's negative connotations of the term. Yet, it is underlined that also the more recent past shaped this negative perception of urban common good, as the new socio-economic system further distorted its image by glorifying individual rights and depreciating the meaning of the common(s). The question 'whose common good?' thus tends to occur in tandem with 'common equals no one's?' (Borowik 2015). Several other authors also pay attention to this condition, reflecting on its detrimental effect on both the present and the future of post-socialist cities (Nawratek 2012, Dymnicka 2013, Grabkowska 2017, 2018).

One of the authors adopts an ideological stance, warning against possible threats following from the current pluralisation of interpretations of the common good by 'researchers and social movements of different character and pedigree' (Słodowa-Hełpa's 2015: 10), which seems to be indicative of subliminal fears of the potential return of 'the communist political and theoretical practice'. The allusion to the 'neomarxist' work of Hardt and Negri (2012), who allegedly attacked the family, the nation-state, and the transnational corporation as 'corrupt forms of the common', is juxtaposed here with Nawratek's (2008) notion of a convergence of traditional forms of social bonds, such as family, with some alternative ones, emerging in physical or digital spaces. Interestingly, Nawratek himself (2012: 105) ponders on the bygone ethos of urban commonality in Poland in relation to the more glorious aspects of the country's past:

> It is striking that in Poland, there are almost no examples of planners or architects representing the interests of local communities against developers and city officials. And yet Poland has a strong tradition of solidarity and community construction, coming from the 1980s. These are not just anecdotes about architects who designed churches with actively participating citizens sitting together at a table in the middle of the village (…) But this tradition of solidarity evaporated long ago under the influence of the 1990s hot money. Today there is nothing left of either.

This rather long inventory of deficiencies held accountable for undermining urban common good is supplemented with a similarly extensive list of measures to be taken. The most widely discussed issue of (the lack of) responsibility and insufficient citizen engagement leads the authors to suggest several remedies. The first one, consisting of the reorientation of education towards the common(s) (Sobol 2016a, Bednarska-Olejniczak and Olejniczak 2018), is regarded as indispensable for the desired socio-cultural change (Grabkowska 2018). The second one lies in the increase of civic participation in decision-making (Dymnicka 2013, Sosnowski 2017). In this case, the solutions vary from general decentralisation and employment of participatory tools (e.g., participatory budgeting) (Sobol 2016b), to fostering bottom-up initiatives and empowerment of local communities (Grabkowska 2018). Another set of recommendations addresses the gaps in the legal framework, aiming at adjusting Polish legislation to meet the challenges of the current revalorisation of common good (Słodowa-Hełpa 2015, Grabkowska 2018).

The Polish city as a commons

An overwhelming majority of the publications linking the notion of commons to the city and urban space build on the writings of Hardin and Ostrom (e.g., Nawratek 2012, Dymnicka 2013, Sadowy 2014, Polko 2015, Słodowa-Hełpa 2015, Bugno-Janik and Janik 2019). The right to the city framework tends to be a requisite in texts published more recently (Borowik 2015, Domaradzka and Wijkström 2016, Grabkowska 2017, 2018, Bugno-Janik and Janik 2019) and is often complemented by Hardt and Negri's idea of the common (Słodowa-Hełpa 2015, Łapniewska 2017, Sokołowicz 2017, Matysek-Imielińska 2018), Foster's and Iaione's city as a commons and Co-city (Łapniewska 2017, Polko 2017, Sokołowicz 2017, Czornik 2018). Related concepts and theories include the prisoner's dilemma and the logic of collective action (Polko 2015), as well as the collective responsibility for common heritage as formulated in the Brundtland Commission (Bugno-Janik et al. 2019).

Although I intended to focus on the holistic conceptualisation of the whole city as a commons, the label of commons is more often assigned to shared, although not necessarily co-managed, fragments or elements of urban space (Sokołowicz 2017, Szewczyk 2018, Wiszniowski 2019). Public space serves as the most common denomination here. For instance, in a paper by Gąsior-Niemiec et al. (2009: 255) the notion appears in relation to strips of no-man's land outside closed residential estates. The authors perceive 'wild or half-wild green areas, local transportation axes—lanes and footpaths, unattended patches of land lying on the outer perimeter of gated communities' either as commons or frontiers, depending on their use and the degree of social divisions between inhabitants of the 'old' and 'new' residential complexes in the neighbourhood.[19] A more conventional approach to viewing public space as a commons is exemplified by Borowik (2015: 96), who operates with a loose criterion of 'universal accessibility and attractiveness for various groups of residents and users'. In contrast, Polko (2015: 172) insists on using more precise economic categories of local public goods and urban commons.

Authors of publications in which the concept of commons is applied to the city as a whole tend to agree that while cities are made up of different types of goods—private, public, club and common-pool—jointly do they constitute an entity on its own (Sobol 2016a sust, Sobol 2016b young, Sadowy 2014). As reckoned by Sadowy (2014: 129), considering the city *en masse* as a common resource is 'all the more important, the more competitive character of the goods that aggregate into it', with investment and consumption pressures reinforcing this effect. Another standpoint, heavily inspired by both the idea of the right to the city concept and a pro-active definition of commons, leans towards commoning. It is adopted by Łapniewska (2017), who frames as such the fight for the right to housing services in Kraków, and for the introduction of participatory budgeting in Sopot. Bugno-Janik and Janik (2019) go even further, seeing commoning as a process which leads to 'development of a system for common management of a range of goods' and, ultimately, enables the transition from the individualist neoliberal urban regime to a collectively oriented one.

Some of the publications dealing with the city as a commons paradigm also raise a number of reservations in terms of the notion's applicability, both in general and with regard to the Polish context. They concern with such issues as indefinite, blurry boundaries (Sadowy 2014), the changeability of users (Sadowy 2014, Czornik 2018), loose community ties, and a high tendency for individualistic behaviours (Bugno-Janik et al. 2019), and the divergence of users' interests and objectives (Sadowy 2014). The practice of urban commoning is therefore presented as more demanding than in the case of non-urban common resources. Yet, as noted by Sadowy (2014), the urban milieu may also facilitate commoning due to, for instance, its prevailing high potential for self-organisation. Her stance is supported by Polko (2015: 176), who refers to the findings of Ostrom in maintaining that cities favour collective action through existing standards of reciprocity, inclusive decision-making, the endogenous potential of users, and the possibility of imposing sanctions.

Another debated issue relates to the identification of actors involved in the making of urban commons. For instance, Czornik (2017) distinguishes six groups of stakeholders: the general population of a city, the local communities, the local authorities, the informal associations, the economic entities, and the supra-local institutions. A complex matrix of interrelations between them supplements the author's proposition of a systematic approach to the problem of creation of urban commons. The heterogeneity of actors is also addressed by Sadowy (2014: 132), who differentiates between users of urban space based on their relationship with the city. On yet another level, the interplay between urban actors is scrutinised using Fligstein and McAdam's theory of fields (2011, 2012) by Domaradzka (2017: 108–109). According to her concept, currently in Poland, new urban movements (*challengers*) have been acting on behalf of the local communities, claiming the urban common good (i.e., public space) from local governments (*incumbents*), with a relevant unit of the central government involved in the process as a third party (*the internal governance unit*). Since there is a possibility for the urban movements to ultimately shift into the position of the incumbents, it is assumed that the counterpublic has the potential to reshape the public sphere.

Several publications provide recommendations championing the development of urban commons in contemporary Polish cities. For instance, Sokołowicz (2017: 38) proposes an assortment of solutions, such as the introduction of adequate legal regulations, fostering new cultural patterns of cooperation and consensus among urban actors, using new technologies in the process of production and management of urban commons, and—interestingly—undermining the domination of an 'anti-common' approach in the public discourse. More practical guidance may be found in the papers by Polko (2015) and Bugno-Janik and Janik (2019), who investigate specific solutions from the opposite ends of the private–public spectrum. The former looks at Business Improvement District (BID) as a tool for collective action in urban areas, both in terms of financing and governance. The latter promotes the idea of an interactive map of urban justice. Based on objective indicators of accessibility to selected urban commons, it is designed to present, for instance, the availability of public transport and public space but also the spatial distribution of physical barriers to disability inclusion, and of local nuisances, such as noise or other forms of pollution.

Finally, a few texts investigate general social attitudes towards the concept of the city as commons. The results of a survey of Polish undergraduates conducted by Sobol (2016a) show a considerable disparity between declarative and actual behaviour. Whereas all of the subjects claim to perceive their city as a common good—most often pointing to public infrastructure (57 per cent), the environment (51 per cent), and public services (39 per cent) as the main categories of urban 'commonality'—only 8 per cent admit to having engaged in any local civic initiatives. Perhaps young age is a major factor here but the notion of the 'crisis of urban commons echoing the crisis of the public' (Sokołowicz 2017: 34) generally rings in the background of most of the analysed publications whenever their authors relate to ongoing changes in the public sphere.

Nevertheless, the urban common(s) narrative in Poland, as well as its incorporation into practice, are considered to be increasingly prevalent. Słodowa-Hełpa (2015: 14) finds it 'optimistic' that the common good had 'entered the contemporary public debate in Poland', whereas Łapniewska (2017: 55) evokes 'the growing number and variety of commons, actively (re)claiming places and spaces'. Papers by Grabkowska (2018) and Bugno-Janik and Janik (2019) evidence the commonality trend in the Polish print media, and the Polish academic discourse. Regardless of occasional pessimism, the outlook is generally affirmative and forward-looking. For example, while Bugno-Janik et al. (2019: 28) find commoning by means of participatory budgeting 'subject to full control of public (municipal) authorities and (…) often steered by them, turning a bottom-up initiative into a manipulated participatory practice', they also refer to it in the context of a 'manifestation of a desired systemic change that may counterbalance the negative consequences of the current, declining and destructive, phase of capitalism'. This transformation is less of a material, than cultural and symbolic character, rendering 'practices in the spirit of commons' a 'symbol (agent) of change'.

Several conclusions may be drawn to sum up this part of the analysis. Firstly, the agglomeration of texts over the analysed period combines with their growing complexity and successively higher understanding of the urban common(s). While the earliest publications are mainly theoretical and dwell on the general idea of the common good in relation to the city and urban issues, there is a clear shift toward more specific conceptualisations of urban commons, and associated practices. Another essential finding is that in all publications the theoretical frameworks derive from Western models, with not much reference to the local knowledge base. Except for two papers relating to the concrete narrative proposed by Mergler et al. (2013), the other ones rely on 'imported' ideas. This, of course, is not in any way inappropriate; however, the lack of attempts to elaborate on the local idiosyncrasies appears to be noteworthy, especially considering the socialist 'tradition' of the paradigm of the city as a communal infrastructure, which provided a frame of reference for urban design and for over four decades. Interestingly, a majority of the investigated academic works do not recall the baggage of the socialist past as a source of valuable experience. Rather, it is either overlooked or treated as unnecessary burden which hampers the expansion of commons in the post-socialist urban milieu.

Yet, the sole fact that the academic debate on urban commoning has evolved from a discussion of the urban common good may be read as a discrete nod to the former period. The potential for inconspicuous inclinations towards the commonality paradigm from the past is twofold, drawing from the then official public sphere, and the counterpublics alike. The former relates to the slogans circulated to cherish the communal urban infrastructure, the latter—to the social teaching of the Catholic Church which encouraged a new way of thinking about 'the common', after the idea had been compromised by the corrupt regime. At the beginning of the 1990s, the concept had the potential for rehabilitation on this ecclesiastical ground, had the former oppositional public sphere morphed into the new official one, but the neoliberal deluge pre-empted any such attempts. So far, the research on commoning in Polish cities has tended to be more theoretically than empirically oriented. However, the well-documented return to the collective logic in terms of urban form and organisation counts as an emerging trend which should not be underrated.

As established in the introduction to this chapter, the changing public discourse is indicative of a wide array of transformations ongoing in Polish cities since the fall of communism. Legal documents, press articles and scholarly works play key roles in setting a discursive frame of reference because of their agenda-setting potential, practical implications, and/or high level of dissemination. What follows from the undertaken discourse analysis is that the recently resurgent interest in urban commonality in Poland has been mostly of bottom-up provenance, and has developed despite relatively adverse top-down processes. The institutional and legal settings which had emerged and evolved during the last three decades did not encourage a communal way of thinking about cities. For instance, the constitutional grounding of the principle of the common good failed to translate into practice. Moreover, post-socialist decentralisation has not led to sufficient empowerment of urban citizens. Even participatory budgeting, a democratic innovation aimed at galvanising local communities into action, has not entirely lived up to expectations. Meanwhile, the 'sanctification' of property rights facilitated the implementation of the neoliberal programme and supported the establishment of the city as a commodity paradigm.

The counter-proposition to this model was first promoted in the media which channelled the right to the city narrative of the rising urban movements. Since around the mid-2010s, with increasing impulses from the academic sphere, the floating signifier of urban common good has taken on new meanings. As they were conveyed to the mainstream and resonated with the public, little by little, the paradigm of the city as a commons consolidated, and effectively clashed with the prevalent vision of the city as a commodity, ultimately eroding the hegemonic neoliberal discourse. The call for overthrowing the anti-commonality orientation of the public debate—postulated explicitly by Sokołowicz (2017: 38)—thus currently loses its radical edge, and becomes an attainable goal.

The centripetal drive towards urban common good has occurred at different paces and scales within the six domains of urban commonality—tending to be most prominent in the commoning of urban space, and less so in practices of urban regeneration. While the results of this part of research are based on public debate, they do not equally represent the voices of all urban stakeholders. Getting to know their perspectives on concrete issues touching upon the question of urban common good in selected Polish cities was the aim of a complementary study, the outcomes of which are presented in Chapter 4.

Notes

1 Among several dividing lines in contemporary Polish society, two stand out as the most distinct. The first one, of a class character, separates the post-gentry intelligentsia elites from the former peasant-worker common people (Giza 2013a). The other, marking a socio-spatial division between Poland A and Poland B, is rooted in the old historical disparity of models of socio-economic development of Polish lands under different rules—the advantaged Prussian one versus the relatively underdeveloped Austrian and Russian ones. This demarcation runs, more or less, along the Vistula and is manifest to this day through diverging demographic, urbanisation, industrialisation, and electoral voting patterns (Zarycki 2013, Harasym et al. 2018).

2 Only between 1989 and 1993 did the government's composition change four times.

3 They are followed by public meetings (75.0 per cent) and digital issues of newspapers (63.5 per cent), the TV (57.7 per cent), the social media (56.1 per cent), the radio (55.9 per cent), blogs dedicated to urban affairs (33.4 per cent), and scientific publications (26.0 per cent).

4 There are three types of gminas in Poland: urban (equivalent to cities or towns), urban–rural (including a city or town and the surrounding non-urban area) and rural (consisting of a non-urban area only). Until 1999 they added up to 49 regions—voivodeships (*województwa*).

5 Other conditions include concern for spatial order, safeguarding architectural and landscape values, natural environment conservation, and public health requirements.

6 It replaced the so-called Small Constitution of 1992 which temporarily regulated the new relations between the legislative and executive powers and, at the same time, acknowledged local government as the 'the basic form of organization of local public life' (Art.70).

7 It was advocated, inter alia, by the Independent Self-Governing Trade Union 'Solidarity' and the constitutional commission of the First-Term Senate. During one of the reading sessions at the National Assembly, Deputy Marshal of the Senate asserted the superiority of 'the man, as a human being with inherent dignity and the resulting inviolable rights' over 'the state that is the common good of all citizens' (Transcript of statements… 1994).

8 In 2015, the revision of the programme introduced the possibility of buying a secondary-market flat. It also allowed singles and unmarried couples to participate in the programme.

9 The other objectives concern preservation of the natural and cultural heritage, resilience in terms of energy and national security, and restoration of spatial order.

10 It needs to be underlined that 'commonality'of several of them is arguable. For instance, the aesthetically pleasing conversions of a former textile factory in Łódź, and a former brewery in Poznań, into shopping centres fall into the category of Privately Owned Public Spaces, whereas a public square redevelopment in Sopot raised controversy among citizens who had wished to have a say at the design phase of the project.

11 Compiled by a media monitoring company, the ranking takes into consideration the number of references to particular media releases appearing in the press, on TV, and on the radio. It thus measures the opinion-forming potential of a media outlet by the frequency with which its content is brought up in other media outlets.

12 In 2004, it had been replaced for the next 15 years by another catchphrase, 'We actually do care' (*Nam nie jest wszystko jedno*). The original motto was eventually restored as a gesture of support for social groups affected by the discriminatory policies of the PiS government: people with disabilities, teachers, the LGBT+ community, doctors, nurses, and judges (*Gazeta Wyborcza*… 2019-07-29).

13 This concerns *Fakt*, *Super Express*, and *Newsweek Polska*.

14 Five of them are included in the present sample (*Gazeta Wyborcza*, *Rzeczpospolita*, *Polityka*, the Polish edition of *Newsweek*, and *Res Publica Nowa*). The remaining sixth was *Dziennik Opinii*, a leftist online daily selected to represent the 'alternative' public sphere, replaced in the current study with a decade younger *Autoportret*.

15 *Rzeczpospolita*'s earliest archives date from 1993, and *Polityka*'s—from 1998. The earliest available online content in *Super Express* and *Newsweek Polska* is from 2001, in *Fakt*—from 2003, and in *Gość Niedzielny*—from 2005. *Res Publica Nowa* and *Autoportet* were digitalised even later than that—in 2008, and 2012, respectively. However, considering their publishing frequency, I was able to investigate print versions of their earlier issues (from 2004 and from 2003, onwards).

16 A parallel text refers to the same group with more common, derogatory denominations of 'lemmings'(known for their desperation and fervour in the battle of life) and 'jars'(derived from their stereotypical reliance on food brought from family homes) (Matys 2015-03-14). Another publicist recapitulates their agenda as 'four times anti'—anti-state, anti-citizenship, anti-solidarity, anti-politics (Makowski 2013-09-16).

17 The Congress was held in 2013 in Gdańsk under the theme 'Common Good—Better Life Quality'.

18 Since BN and CEEOL allow searching not only by content but by keywords as well, the procedure was repeated to include both options. The much narrower search by keywords helped to eliminate the 'false' results which often popped up in searches by content. Some of them would be thematically unrelated to the issues under study, but the search engine recognised them as valid because of their creative commons licenses or use of Wikimedia commons media.

19 The paradoxical connection between gating and its antithesis—commoning, is studied in a paper by Grabkowska and Szmytkowska (2021), which had appeared in the search results but did not become part of the analysis due to the publication date.

References

(2000-10-14) Płot z ostrym szpicem. *Polityka*. Retrieved from https://www.polityka.pl/archiwumpolityki/1849824,1,8222plot-z-ostrym-szpicem8221.read

(2003-07-30) Miasto—zbiorowy obowiązek. *Gazeta Wyborcza*. Retrieved from https://classic.wyborcza.pl/archiwumGW/2154484/MIASTO---ZBIOROWY-OBOWIAZEK

(2003-09-16) Z troską o miastach. *Gazeta Wyborcza*. Retrieved from https://classic.wyborcza.pl/archiwumGW/2193848/Z-TROSKA-O-MIASTACH

(2004-12-11) Ciemność widzę. *Polityka*. Retrieved from https://www.polityka.pl/archiwumpolityki/1832648,1,ciemnosc-widze.read

(2011-11-24) Prezes bogacz ukrywa dochody. *Fakt*. Retrieved from https://www.fakt.pl/wydarzenia/polska/prezes-bogacz-ukrywa-dochody/cr71gzv

(2014-01-02) We własnych rękach. *Gość Niedzielny*. Retrieved from https://www.gosc.pl/doc/1828195.We-wlasnych-rekach

(2018-06-04) Chcemy otwierać oczy na architekturę. *Polityka*. Retrieved from https://www.polityka.pl/tygodnikpolityka/ludzieistyle/1751086,1,chcemy-otwierac-oczy-na-architekture.read?page=7&moduleId=4721

Act on the Amendment of Certain Acts in Connection with the Strengthening of Landscape Protection Tools of 24 April 2015 (2015) Journal of Laws, item 774.

Act the Amendment of Certain Laws to Increase Citizens' Participation in the Process of Election, Operation and Control of Some Public Bodies of 11 January 2018 (2018) Journal of Laws, item 130.

Act on the Amendment of the Act on Nature Conservation and to the Act on Forests of 16 December 2016 (2016) Journal of Laws, item 2149.
Act on Commune Self-Government of 8 March 1990 (1990) Journal of Laws no. 16, item 95.
Act on the Direct Election of a Commune Head, Town Mayor and President (2002) Journal of Laws no. 113, item 984.
Act on the Introduction of the Three-tier Division of the Country of 24 July 1998 (1998) Journal of Laws no. 98, item 603.
Act on Nature Conservation of 16 April 2004 (2004) Journal of Laws no. 92, item 880.
Act on Powiat Self-Government of 5 June 1998 (1998) Journal of Laws no. 91, item 578.
Act on Collective Public Transport of 16 December 2010 (2011) Journal of Laws no. 5, item 13.
Act on Revitalisation of 9 October 2015 (2015) Journal of Laws, item 177.
Act on Spatial Development of 7 July 1994 (1994) Journal of Laws no. 15, item 139.
Act on Protection and Care of Monuments of 27 July 2003 (2003) Journal of Laws no. 162, item 1568.
Act on Spatial Planning and Development of 27 March 2003 (2003) Journal of Laws no. 80, item 717.
Act on Voivodeship Self-Government of 5 June 1998 (1998) Journal of Laws no. 91, item 576.
April Constitution of Poland (1935) Journal of Laws no. 30, item. 227.
Adamczyk S (2015-11-12) Nielegalne rabatki zamiast śmieci. *Super Express*. Retrieved from https://www.se.pl/krakow/nielegalne-rabatki-zamiast-smieci-audio-wideo-aa-soTC-rywu-wpeQ.html
Arnstein SR (1969) A ladder of citizen participation. *JAIP*, 35(4), 216–224.
Asen R (2017) Neoliberalism, the public sphere, and a public good. *Quarterly Journal of Speech*, 103(4), 329–349. https://doi.org/10.1080/00335630.2017.1360507
Asman T, Niewiadomski Z (2018) Przepisy ogólne. In: Niewiadomski Z (ed.), *Prawo budowlane: Komentarz*. Warszawa: Wydawnictwo CH Beck, 3–178.
Bartoszewicz D, Jarecka D (2008-07-19) Do szczęścia potrzebne jest miasto: Interview with Magdalena Staniszkis, *Gazeta Wyborcza*. Retrieved from https://classic.wyborcza.pl/archiwumGW/5143588/Do-szczescia-potrzebne-jest-miasto
Bednarska-Olejniczak D, Olejniczak J (2018) Problem partycypacji obywateli w kształtowaniu działań gminy. *Przedsiębiorczość i Zarządzanie*, 19(9/1), 129–144.
Beim M (2011-05-07) Raport z zakorkowanego miasta. *Rzeczpospolita*. Retrieved from http://archiwum.rp.pl/1045968.html
Bendyk E (2017-12-26) Walka o samorządy dopiero się zacznie. *Polityka*. Retrieved fromhttps://www.polityka.pl/tygodnikpolityka/kraj/1732234,1,walka-o-samorzady-dopiero-sie-zacznie.read
Bielecki C (2012-01-14) Chaos i zgiełk. *Rzeczpospolita*. Retrieved from https://archiwum.rp.pl/1112266.html
Bielecki C (2015-09-05) Prawo do piękna. *Rzeczpospolita*. Retrieved from http://archiwum.rp.pl/1285411.html
Bierwiaczonek K (2016) *Społeczne wytwarzanie przestrzeni publicznych*. Katowice: Wydawnictwo Uniwersytetu Śląskiego.
Bierwiaczonek K, Dymnicka M, Kajdanek K, Nawrocki T (2017) *Miasto, przestrzeń, tożsamość: Studium trzech miast: Gdańsk, Gliwice, Wrocław*. Warszawa: Wydawnictwo Naukowe Scholar.
Bittner K (2007-08-17) Dotleniacz dla wszystkich. *Gazeta Wyborcza*. Retrieved from https://classic.wyborcza.pl/archiwumGW/4927459/Dotleniacz-dla-wszystkich

Bodei S (2015) Nowe modele mieszkania i wspólnoty w Barcelonie. *Autoportret*, 4(51), 22–28.

Borowik I (2015) Władze miejskie a poszukiwania dobrej przestrzeni publicznej. *Acta Universitatis Lodziensis: Folia Sociologica*, 54, 95–108. http://dx.doi.org/10.18778/0208-600X.54.07

Bogucka T (1996-05-10) Nasza bezradnoŚć, nasza zaradnoŚć: Interview with Janusz Grzelak. *Gazeta Wyborcza*. Retrieved from https://classic.wyborcza.pl/archiwumGW/214565/Nasza-bezradnosc--nasza-zaradnosc

Brzeziński M (2018-12-19) Poznaniacy nie dbają o dobro wspólne? Hulajnogi są dosłownie wszędzie i psują krajobraz. *Super Express*. Retrieved from https://www.se.pl/poznan/poznaniacy-nie-dbaja-o-dobro-wspolne-hulajnogi-sa-doslownie-wszedzie-i-psuja-krajobraz-aa-4UuT-Lx2z-sNKB.html

Bugno-Janik A, Cymbrowski B, Janik M, Łuksza J (2019) Urban commons: czy z utopii można zrobić rzeczywistoŚć? *Kultura Współczesna*, 3(106), 26–39. http://doi.org/10.26112/kw.2019.106.03

Bugno-Janik A, Janik M (2019) Miasto sprawiedliwe w praktyce. *GórnoŚląskie Studia Socjologiczne*, 1, 159–177.

Bukowiecki Ł, Obarska M, Stańczyk X (eds) (2014) *Miasto na żądanie. Aktywizm, polityki miejskie*. Warszawa: Wydawnictwa Uniwersytetu Warszawskiego.

Buzek J (2001-08-30) Samorządni - znaczy wolni. *Gazeta Wyborcza*. Retrieved from https://classic.wyborcza.pl/archiwumGW/1519122/SAMORZADNI---ZNACZY-WOLNI

Cheng L, Danesi M (2019) Exploring legal discourse: a sociosemiotic (re)construction. *Social Semiotics*, 29(3), 279–285. http://doi.org/10.1080/10350330.2019.1587841

Constitution of the Polish People's Republic (1952) Journal of Laws no. 33, item 232.

Constitution of the Republic of Poland (1997) Journal of Laws no. 78, item 483.

Construction Law (1994) Journal of Laws no. 89, item 414.

Czapiński J (2015) Stan społeczeństwa obywatelskiego. In: Czapiński J, Panek T (eds) (2015) Social Diagnosis 2015: Objective and subjective quality of life in Poland. *Contemporary Economics*, 9(4), 332–372.

Czapiński J, Sułek A (2011) Stan społeczeństwa obywatelskiego. In: Czapiński J, Panek T (eds) Social Diagnosis 2011: Objective and subjective quality of life in Poland. *Contemporary Economics*, 5(3), 271–298.

Czarnecki B (2017) Konflikty i różnice interesów w zagospodarowaniu przestrzeni a dobro wspólne w systemie planowania przestrzennego. In: Banaszuk P, Tokajuk J (eds), *Problemy planowania przestrzennego: Podlaskie Forum Urbanistów 2012-2016*. 78–87.

Czech S, Kassner M (2021) Counter-movement at a critical juncture: a neo-Polanyian interpretation of the rise of the illiberal Right in Poland. *Intersections. EEJSP*, 7(2), 128–148. https://doi.org/10.17356/ieejsp.v7i2.733

Czornik M (2017) Twórcy miejskich dóbr wspólnych. *Studia Miejskie*, 28, 43–58.

Czornik M (2018) MiejskoŚć dóbr wspólnych: Refleksje nad adaptowaniem koncepcji Common-Pool Resources. *Rozwój Regionalny i Polityka Regionalna*, 43, 71–82.

Czyńska M (2010-11-13) Alternatywy 12. *Gazeta Wyborcza*. Retrieved from http://www.archiwum.wyborcza.pl/Archiwum/2029040,0,7311019,20101113RP-TOB,ALTERNATYWY_12,zwykly.html

Democracy Declining: Erosion of Media Freedom in Poland MEDIA FREEDOM RAPID RESPONSE (MFRR) PRESS FREEDOM MISSION TO POLAND (November-December 2020) MISSION REPORT https://ipi.media/wp-content/uploads/2021/02/20210211_Poland_PF_Mission_Report_ENG_final.pdf (last accessed on 14 November 2021)

Dobek-Ostrowska B (2019) How the media systems work in Central and Eastern Europe. In: Połońska E, Beckett C (eds), *Public Service Broadcasting and Media Systems in Troubled European Democracies*, 259–278, Cham: Palgrave Macmillan. https://doi.org/10.1007/978-3-030-02710-0_12

Domaradzka A (2017) Leveling the playfield: Urban movement in the strategic action field of urban policy in Poland. In: Hou J, Knierbein S (eds), *City Unsilenced: Urban Resistance and Public Space in the Age of Shrinking*. 106–118.

Domaradzka A, Wijkström F (2016) Game of the city re-negotiated: the polish urban re-generation movement as an emerging actor in a strategic action field. *Polish Sociological Review*, 3(195), 291–308.

Domaradzka A (2021) *Klucze do miasta: Klucze do miasta Ruch miejski jako nowy aktor w polu polityki miejskiej. Warszwa*: Scholar.

Drozda Ł (2016) WłasnoŚć w modelu neoliberalnym na przykładzie polskiej przestrzeni zurbanizowanej po 1989 roku. *Studia Regionalne i Lokalne*, 4(66), 48–61. https://doi.org/10.7366/1509499546603

Dudek J (2006-03-13) Wolnoć Tomku… *Rzeczpospolita*. Retrieved from https://archiwum.rp.pl/604094.html

Dymnicka M (2013) *Przestrzeń publiczna a przemiany miasta*. Warszawa: Wydawnictwo Naukowe Scholar.

Erbel J (2020) *Poza własnością: W stronę udanej polityki mieszkaniowej*. Kraków: Wysoki Zamek.

Erbel J, Ratajczak M (2013). Prawo do dobra wspólnego. *Miasta*, 1(2), 38–43.

European Charter of Local Self-Government (2014) Journal of Laws no. 124, item 607.

Ferenčuhová S (2012) Urban theory beyond the 'East/West divide'? Cities and urban research in postsocialist Europe. In: T Edensor, M Jayne (eds.), *Urban Theory beyond the West: A World of Cities*. London and New York: Routledge, 65–74.

Filas R, Płaneta P (2009) Media in Poland and public discourse. In: Czepek A, Hellwig M, Nowak E (eds) *Press Freedom and Pluralism in Europe: Concepts and Conditions*. Bristol, UK: Intellect Books, 141–163.

Fligstein N, McAdam D (2011) *A Theory of Fields*. New York: Oxford University Press

Fligstein N, McAdam D (2012) Social Skill and the Theory of Fields. *Sociological Theory*, 19(2), 105–125.

Gąsior-Niemiec A, Glasze G, Pütz R (2009) A Glimpse over the Rising Walls: The Reflection of Post-Communist Transformation in the Polish Discourse of Gated Communities. *East European Politics and Societies*, 23, 244–265. https://doi.org/10.1177/0888325408328749

Gdula M (2013-07-25) Jak dobrze, że młodzi znów zawiedli. *Gazeta Wyborcza*. Retrieved from https://classic.wyborcza.pl/archiwumGW/7767427/Jak-dobrze--ze-mlodzi-znow-zawiedli

Gehl J (2009) *Życie między budynkami: Użytkowanie przestrzeni publicznych*. Translated by Urbańska MA, Wydawnictwo RAM: Kraków.

Gendźwiłł A, Swianiewicz P (2017) Breeding grounds for local independents, bonus for incumbents: Directly elected mayors in Poland. In: Sweeting D (ed.), *Directly Elected Mayors in Urban Governance: Impact and Practice*. Policy Press, 179–200. https://doi.org/10.1332/policypress/9781447327011.003.0011

Giedrys G (2014-03-29) Pani sprzątająca po psie na Wawel! Interview with Tomasz Szlendak. *Gazeta Wyborcza*. Retrieved from https://classic.wyborcza.pl/archiwumGW/7864472/Pani-sprzatajaca-po-psie-na-Wawel-

Gitkiewicz O (2019) *Nie zdążę*. Warszawa: Dowody na Istnienie.

Giza A (2013a) Dwie Polski o dwóch Polskach, czyli samoreprodukujący się dyskurs. In: Giza A, Bekas P, Darmas M, Dembek A, Drogowska K, Gołdys A, Halawa M, Lewicki M, Marody M, Nowotny A, Strzemińska A, Wasilewski J, *Gabinet luster: o kształtowaniu samowiedzy Polaków w dyskursie publicznym.* Warszawa: Wydawnictwo Naukowe Scholar, 105–145.

Giza A (2013b) Wprowadzenie. O nowych szatach cesarza, wiedzy wspólneji badaniu samowiedzy społecznej. In: Giza A, Bekas P, Darmas M, Dembek A, Drogowska K, Gołdys A, Halawa M, Lewicki M, Marody M, Nowotny A, Strzemińska A, Wasilewski J, (eds) *Gabinet luster: O kształtowaniu samowiedzy Polaków w dyskursie publicznym.* Warszawa: Wydawnictwo Naukowe Scholar, 9–31.

Gontarz J, Gutowska I (eds) (2014) *Wspólna przestrzeń—wspólne dobro. Dobre praktyki w kształtowaniu ładu przestrzennego.* Warszawa: Ministerstwo Infrastruktury i Rozwoju.

Goodnight GT (1987) Public discourse. *Critical Studies in Mass Communication*, 4(4), 4428–432. https://doi.org/10.1080/15295038709360154

Governance Act of 3 May (1791) Retrieved from: https://en.wikisource.org/wiki/Constitution_of_3_May_1791

Grabkowska M (2017) Przestrzeń miasta postsocjalistycznego jako dobro wspólne. *Przegląd koncepcji teoretycznych, Prace Geograficzne*, 149, 33–52.

Grabkowska M (2018) Urban space as a commons in print media discourse in Poland after 1989. *Cities*, 71, 22–29.

Grabkowska M, Szmytkowska M (2021) Gating as Exclusionary Commoning in a Post-socialist City. *Region*, 8(1), 15–32.

Gusenbauer M, Haddaway NR (2020). Which academic search systems are suitable for systematic reviews or meta-analyses? Evaluating retrieval qualities of Google Scholar, PubMed, and 26 other resources. *Research Synthesis Methods*, 11(2), 181–217. https://doi.org/10.1002/jrsm.1378

Gzowska A (2019) Urbanista w konflikcie: Interview with Paweł Jaworski. *Autoportret*, 1(64), 20–26.

Hac A (2014-08-28) Miasta budzą się do życia. Gazeta Wyborcza. Retrieved from: http://www.archiwum.wyborcza.pl/Archiwum/2029040,0,7922878,20140828RP-DGW,Miasta_budza_sie_do_zycia,.html

Harasym R, Pater R, Skica T (2018) Konkurencyjność i rozwój Polski Wschodniej. *Samorząd Terytorialny*, 5, 64–76.

Hardt M, Negri A (2012) *Rzecz-pospolita. Poza własność prywatną i dobro publiczne.* Translated by Juskowiak P, Kowalczyk A, Ratajczak M, Szadkowski K, Szlinder M, Kraków: Korporacja Ha!art.

Harvey D (2012) *Bunt miast: Prawo do miasta i miejska rewolucja.* Translated by Kowalczyk A, Marzec W, Mikulewicz M, Szlinder M, Warszawa: Fundacja Nowej Kultury Bęc Zmiana.

Hawranek M (2014-01-16) Café bez pieniędzy *Gazeta Wyborcza.* Retrieved from http://www.archiwum.wyborcza.pl/Archiwum/2029040,0,7834624,20140116RP-TDF,Caf_bez_pieniedzy,.html

Hołownia S (2011-03-12) Kupa polska. *Newsweek Polska.* Retrieved from https://www.newsweek.pl/kupa-polska/4lx7xwt

Jacobs J (2017) *Śmierć i życie wielkich miast Ameryki.* Translated by Mojsak Ł, Warszawa: Fundacja Nowej Kultury Bęc Zmiana.

Jacobs K (2006) Discourse Analysis and its Utility for Urban Policy Research. *Urban Policy Research*, 24(1), 39–52.

Jacobsson K (2017) Rethinking civic privatism in a postsocialist context: Individualism and personalization in polish civil society organizations. In: Jacobsson K, Korolczuk E (eds), *Civil Society Revisited: Lessons from Poland*. New York: Berghahn, 81–104.

Jacobsson K, Korolczuk E (eds.) (2017) *Civil Society Revisited: Lessons from Poland*. New York: Berghahn Books, 176–199.

Jałoszewski M, Paś W (2015-04-13) Prawo własności to żadne sacrum. *Gazeta Wyborcza*. Retrieved from https://classic.wyborcza.pl/archiwumGW/8009694/PRAWO-WLASNOSCI-TO-ZADNE-SACRUM

Janicki M (2002-03-23) Partyjna grządka. *Polityka*. Retrieved from https://www.polityka.pl/archiwumpolityki/1808216,1,partyjna-grzadka.read

Jarkowiec M (2016-10-15) Czy można oddać miasto ludziom. *Gazeta Wyborcza*. Retrieved from https://classic.wyborcza.pl/archiwumGW/8205039/CZY-MOZNA-ODDAC-MIASTO-LUDZIOM

Jaworski R (2004-10-12) Chaos przestrzenny. *Gazeta Wyborcza*. Retrieved from: https://classic.wyborcza.pl/archiwumGW/644410/Tylko-bez-samowoli

Jedlecki P (2007-10-01) Małe jest piękne, ale duży może więcej: Interview with Marek Szczepański. *Gazeta Wyborcza*. Retrieved from https://classic.wyborcza.pl/archiwumGW/4949239/Male-jest-piekne--ale-duzy-moze-wiecej

Jędraszko A (2005) *Zagospodarowanie przestrzenne w Polsce – drogi i bezdroża regulacji ustawowych*. Warszawa: Unia Metropolii Polskich.

Kacprzak L, Koszel B, Marcinkowski A (eds) (2012) *Społeczeństwo obywatelskie jako dobro wspólne*. Piła: Państwowa Wyższa Szkoła Zawodowa im. *Stanisława Staszica w Pile, Piła* 2012, 137–151.

Kamiński AZ (2008-05-28) Z której strony psuje się ryba. *Rzeczpospolita*. Retrieved from http://archiwum.rp.pl/779733.html

Karwowska A (2019-02-20) Dobrze żyje się tam, gdzie prezydent ma osobowość: Interview with Grzegorz Gorzelak. *Gazeta Wyborcza*. Retrieved from https://classic.wyborcza.pl/archiwumGW/9021532/DOBRZE-ZYJE-SIE-TAM--GDZIE-PREZYDENT-MA-OSOBOWOSC

Kasiński M (2018) Ethical and political dilemmas of local self-government in Poland in the course of systemic transformations (1990–2018). *Annales Ethics in Economic Life*, 21(7), 7–26. http://dx.doi.org/10.18778/1899-2226.21.7.01

Kipta E (2015-06-01) Gwarancje dla dóbr wspólnych. *Rzeczpospolita*. Retrieved from http://archiwum.rp.pl/1277839.html

Klich A (2011-06-04) Wyrastamy ze wsi: Interview with Paweł Kubicki. *Gazeta Wyborcza*. Retrieved from https://classic.wyborcza.pl/archiwumGW/7421527/Wyrastamy-ze-wsi

Kmieciak S (2018-01-25) Budżet obywatelski: najwyższy czas na wersję 2.0? *Gazeta Wyborcza*. Retrieved from: https://classic.wyborcza.pl/archiwumGW/8317252/BUDZET-OBYWATELSKI--NAJWYZSZY-CZAS-NA-WERSJE-2-0-

Koczanowicz L (2016) The Polish Case: Community and Democracy under the PiS. *New Left Review*, 102, 77–96.

Korcelli-Olejniczak E, Bierzyński A, Dworzański P, Grochowski M, Piotrowski F, Węcławowicz G (2017). *DIVERCITIES: Dealing with Urban Diversity. The Case of Warsaw*. Warsaw: IGSO PASc.

Korolczuk E (2017) When parents become activists: Exploring the intersection of civil society and family. In: Jacobsson K, Korolczuk E (eds), *Civil Society Revisited: Lessons from Poland*. New York: Berghahn, 129–152.

Kotus J, Sowada T, Rzeszewski M, Mańkowska P (2019) Anatomy of place-making in the context of the communication processes: A story of one community and one square in a post-socialist city. *Quaestiones Geographicae*, 38(2), 51–66.

Kompowski A (2016-10-21) Miasto dla ludzi, nie tylko dla aut: Interview with Hubert Igliński. *Gazeta Wyborcza*. Retrieved from https://classic.wyborcza.pl/archiwumGW/8206713/MIASTO-DLA-LUDZI--NIE-TYLKO-DLA-AUT

Kowalewski A (2010-07-21) Problem z ładem przestrzennym. *Rzeczpospolita*. Retrieved from http://archiwum.rp.pl/964369.html

Kowaluk M (2011) Public space in contemporary housing estates – attention to the common good versus maximization of investor's income. *Architectus*, 2(30), 71–72.

Koźlenko D (2014-08-19) Letnie obyczaje Polaków. *Polityka*. Retrieved from https://www.polityka.pl/tygodnikpolityka/kraj/1589603,1,letnie-obyczaje-polakow.read

Krier L (2011) *Architektura wspólnoty*. Gdańsk: słowo/obraz terytoria.

Kronenberg J, Łaszkiewicz E, Sziło J (2021) Voting with one's chainsaw: What happens when people are given the opportunity to freely remove urban trees? *Landscape and Urban Planning*, 209, 104041, https://doi.org/10.1016/j.landurbplan.2021.104041

Krupa-Dąbrowska R (2013a-06-10) Reklamy da się ujarzmić. Interview with Olgierd Dziekoński. *Rzeczpospolita*. Retrieved from. http://archiwum.rp.pl/1191438.html

Krupa-Dąbrowska R (2013b-11-22) Lokator musi usunąć brzydką stolarkę. *Rzeczpospolita*. Retrieved from http://archiwum.rp.pl/1224748.html

Krysiak P (2019-04-23) Politycy, nie przeszkadzajcie nauczycielom. *Gazeta Wyborcza*. Retrieved from https://classic.wyborcza.pl/archiwumGW/9033485/POLITYCY--NIE-PRZESZKADZAJCIE-NAUCZYCIELOM

Kubicki P (2016) *Wynajdywanie miejskości*. Kraków: Nomos.

Kubicki P (2020) Miasto jako dobro wspólne czy suma prywatnych własności? O polskiej kwestii miejskiej w kontekście działalności ruchów miejskich. *Przegląd Kulturoznawczy*, 4(46), 477–492.

Kurdupski M (2015-08-03) "Wprost" o 72 proc. w dół, a "Polityka" - o 32 proc. Sprzedaż tygodników opinii od 2005 roku (raport), https://www.wirtualnemedia.pl/artykul/wprost-o-72-proc-w-dol-a-polityka-o-32-proc-sprzedaz-tygodnikow-opinii-od-2005-roku-raport

Lees L (2004) Urban geography: discourse analysis and urban research. *Progress in Human Geography*, 28(1), 101–107.

Legutko P (2013-08-29) Mała solidarność. *Gość Niedzielny*. Retrieved from https://www.gosc.pl/doc/1681558.Mala-solidarnosc/2

Leszczyński A (2006-08-12) Polak na zagrodzie. *Gazeta Wyborcza*. Retrieved from https://classic.wyborcza.pl/archiwumGW/4695339/POLAK-NA-ZAGRODZIE

Leśniak-Rychlak D (2018/2019) *Jesteśmy wreszcie we własnym domu*. Kraków: Instytut Architektury.

Leśniak-Rychlak D (2012) Nadzieja na uczestnictwo: Interview with Krzysztof Nawratek. *Autoportret*, 2(37), 45–47.

Lichocka J (2010-07-20) Jak trafić do serc nowych wielkomiejskich. *Rzeczpospolita*. Retrieved from http://archiwum.rp.pl/964219.html

Lipczak A (2014-04-12) Spółdzielnia. *Gazeta Wyborcza*. Retrieved from http://www.archiwum.wyborcza.pl/Archiwum/2029040,0,7867401,20140412RP-TWY,SPOLDZIELNIA,.html

Lipoński S (2013-11-08) Obywatele sami dzielą miejskie pieniądze: Interview with Jarosław Makowski. *Gazeta Wyborcza.* Retrieved from https://classic.wyborcza.pl/archiwumGW/7810108/OBYWATELE-SAMI-DZIELA-MIEJSKIE-PIENIADZE

Luczys P (2011) Miasto jako tekst – odzyskiwanie przestrzeni dyskursu. In: Nowak M, Pluciński P (eds), *O miejskiej sferze publicznej. Obywatelskość i konflikty o przestrzeń.* Kraków: Korporacja Ha!art, 283–294.

Łabendowicz O (2018) Dyskursywna ewolucja wyrażenia dobra zmiana. *Acta Universitatis Lodziensis Folia Linguistica*, 52, 163–178. http://dx.doi.org/10.18778/0208-6077.52.12

Łapniewska Z (2017) (Re)claiming Space by Urban Commons. *Review of Radical Political Economics*, 49(1), 54–66.

Łazarczyk G (2018-07-17) "Mieszkanie Plus" jak za Gierka: interview with Artur Soboń. *Gazeta Wyborcza.* Retrieved from https://classic.wyborcza.pl/archiwumGW/8357015/-MIESZKANIE-PLUS--JAK-ZA-GIERKA

Łazarczyk G (2019-01-30) Metropolie nie "wysysają" ludzi: Interview with Grzegorz Gorzelak. *Gazeta Wyborcza.* Retrieved from https://classic.wyborcza.pl/archiwumGW/9017462/Metropolie-nie--wysysaja--ludzi

Machocka M, Śwital P (2016) Dobro wspólne jako czynnik determinujący działanie gminy. In: Kułak-Krzysiak K, Parchomiuk J (eds), *Służąc dobru wspólnemu.* Lublin: Wydawnictwo KUL, 135–150.

Maciejewski J (2017-03-11) Dobro wspólne wraca oknem. *Rzeczpospolita.* Retrieved from https://archiwum.rp.pl/artykul/1335980-Dobro-wspolne-wraca-oknem.html

Majcherek JA (2011-08-22) Z laptopem w obdrapanych wagonach. *Gazeta Wyborcza.* Retrieved from https://classic.wyborcza.pl/archiwumGW/7460782/Z-laptopem-w-obdrapanych-wagonach

Makowski J (2013-09-16) Mieszczanie na ławie oskarżonych. *Rzeczpospolita.* Retrieved from http://archiwum.rp.pl/1217562.html

Malinowska A (2010-02-17) Straż parkowa. *Gazeta Wyborcza.* Retrieved from https://classic.wyborcza.pl/archiwumGW/7166134/Straz-parkowa

Markowski MP (2019) *Wojny nowoczesnych plemion. Spór o rzeczywistość w epoce populizmu.* Kraków: Karakter.

Matys M (2015-03-14) Waćpan, co sobą jest zaćpan. *Gazeta Wyborcza.* Retrieved from https://classic.wyborcza.pl/archiwumGW/7998970/Wacpan--co-soba-jest-zacpan

Matysek-Imielińska M (2018) *Miasto w działaniu. Warszawska Spółdzielnia Mieszkaniowa - dobro wspólne w epoce nowoczesności.* Warszawa: Fundacja Bęc Zmiana.

Maziarski W (2015-02-19) Nie podważajmy instytucji własności. *Gazeta Wyborcza.* Retrieved from https://classic.wyborcza.pl/archiwumGW/7989616/NIE-PODWAZAJMY-INSTYTUCJI-WLASNOSCI

Mazur N (2015-07-23) Na szczęście zawsze coś się zawali: Interview with Jakub Głaz. *Gazeta Wyborcza.* Retrieved from https://classic.wyborcza.pl/archiwumGW/8045342/NA-SZCZESCIE-ZAWSZE-SIE-COS-ZAWALI

Mączka K, Jeran A, Matczak P, Milewicz M, Allegretti G (2021) Models of participatory budgeting. Analysis of participatory budgeting procedures in Poland. *Polish Sociological Review*, 216(4), 473–492.

McCombs ME, Shaw DL (1972) The agenda-setting function of mass media. *Public Opinion Quarterly*, 36, 176–185. https://doi.org/10.1086/267990

McQuire S (2008) *The Media City: Media, Architecture and Urban Space*. London: Sage.

Mencwel, J (2020) *Betonoza:. Jak się niszczy polskie miasta*. Warszawa: Wydawnictwo Krytyki Politycznej 2020.

Mergler L (2012) Sukces, kryzys i bunt poznańskich mieszczan. *Res Publica Nowa*, 208(18), 43–45.

Mergler L, Pobłocki K (2010) Nic o nas bez nas: polityka skali a demokracja miejska. *Res Publica Nowa*, 201–202(11–12), 7–14.

Mergler L, Pobłocki K, Wudarski M (2013) *Anty-Bezradnik przestrzenny: prawo do miasta w działaniu*. Warszawa: Fundacja Res Publica im. H. Krzeczkowskiego.

Micek M, Staszewska S (2019) Urban and Rural Public Spaces: Development Issues and Qualitative Assessment. *Bulletin of Geography. Socio-economic Series*, 45(45), 75–93. http://doi.org/10.2478/bog-2019-0025

Milarczyk A (2019) Budowanie tożsamości organizacyjnej – studium przypadku dziennika „Rzeczpospolita". *Zarządzanie Mediami*, 7(2), 97–115. http://doi.org/10.4467/23540214ZM.19.007.10929

Millard F (2008) Party politics in Poland after the 2005 elections. In: Myant M, Cox T (eds.) *Reinventing Poland: Economic and political transformation and evolving national identity*. Abingdon: Routledge, 65–82.

Milewska P, Pacewicz K (2018-09-21) Nadzieja w ruchach miejskich. *Gazeta Wyborcza*. Retrieved from https://classic.wyborcza.pl/archiwumGW/8371632/NADZIEJA-W-RUCHACH-MIEJSKICH

Montgomery C (2015) *Miasto szczęśliwe: Jak zmienić nasze życie, zmieniając nasze miasta*. Kraków: Wysoki Zamek.

Mucha J (2017-10-12) Wspólnota, praca, szkoła... Naprawmy je razem. https://classic.wyborcza.pl/archiwumGW/8294749/WSPOLNOTA--PRACA--SZKOLA----NAPRAWMY-JE-RAZEM

Müller M (2019) Goodbye, Postsocialism! *Europe-Asia Studies*, 71(4), 533–550. http://doi.org/10.1080/09668136.2019.1578337

National Spatial Development Concept 2030 (2012) Official Gazette of the Republic of Poland, item 252.

National Urban Policy 2023 (2015) Official Gazette of the Republic of Poland, item 1235.

Nawratek K (2008) Miasto jako idea polityczna, Korporacja Ha!art, Kraków.

Nawratek K (2012) *Holes in the Whole*.

Nizinkiewicz J (2013-08-30) Wrocław: historia przezwyciężona. *Rzeczpospolita*. Retrieved from http://archiwum.rp.pl/1215997.html

Nowak M (2010) Strukturalne uwarunkowania gry o przestrzeń. Na przykładzie poznańskiego projektu: „Karta przestrzeni publicznej". In: Kryczka P, Bielecka-Prus J (eds.), *Przemiany miast polskich po 1989*. Lublin: Wydawnictwo WSPA, 125–149.

Nowakowski M (2009-06-13) Czy odrodzą się małe wspólnoty obywatelskie? *Rzeczpospolita*. Retrieved from http://archiwum.rp.pl/871891.html

Nowotarski B (2014-11-13) Oddać samorządy obywatelom. *Gazeta Wyborcza*. Retrieved from https://classic.wyborcza.pl/archiwumGW/7953381/ODDAC-SAMORZADY-OBYWATELOM

Pacholski A (2014-03-08) Samochodoza polonika. *Gazeta Wyborcza*. Retrieved from https://classic.wyborcza.pl/archiwumGW/7855890/Samochodoza-polonika

Pancewicz Ł (2013) Miasto jako dobro wspólne. Od „użytkowników" miasta do obywateli. In: Pancewicz Ł, Zbieranek P (eds), *Pomorskie miasta–jak je kształtować dla dobra wspólnego?* Gdańsk: Instytut Badan nad Gospodarką Rynkową, 13–17.

Peterson T, Jensen JM, Rivers WR (1966) *The Mass Media and Modern Society*. New York: Holt, Rinehart & Winston.

Piechowiak M (2012) *Dobro wspólne jako fundament polskiego porządku konstytucyjnego*. Warszawa: Biuro Trybunału Konstytucyjnego.

Pistelok P, Martela B (eds) (2019) Raport o stanie polskich miast: Partycypacja publiczna.

Plichta M (2015-10-30) Tam było robione coś, tylko nic się nie uchowało. *Autoportret*. Retrieved from https://autoportret.pl/tam-bylo-robione-cos-tylko-nic-sie-nie-uchowalo/

Pluciński P (2018) Forces of altermodernization: Urban social movements and the new urban question in contemporary Poland. *Voluntas*, 29, 653–669. https://doi.org/10.1007/s11266-018-0007-x

Pobłocki K (2012-10-19) Magma ruchów miejskich. *Res Publica Nowa*. Retrieved from https://publica.pl/teksty/magma-ruchow-miejskich-32603.html

Pobłocki K (2015-05-20) Optyczna większoŚć. *Res Publica Nowa*. Retrieved from: https://publica.pl/teksty/optyczna-wiekszosc-51653.html

Podgórska J (2007-01-06) Trzecie. *Polityka*. Retrieved from https://www.polityka.pl/tygodnikpolityka/klasykipolityki/1802268,1,jak-oswoic-miasto.read

Polko A (2015) Strategie kolektywnego działania w miejskich przestrzeniach publicznych. *Studia KPZK PAN*, 161, 172–179.

Polko A (2017) *Miasto jako dobro wspólne: Czy zamierzamy w kierunku wspólnoty współpracy i gospodarki współdzielenia?* In: Drobniak A (ed.), Nowe sektory gospodarki w rozwoju miasta - hybrydyzacja rozwoju. Katowice: Uniwersytet Ekonomiczny w Katowicach, 117–130.

Polska 2030 - Trzecia fala nowoczesnoŚci: Długookresowa Strategia Rozwoju Kraju (2013) Warszawa: Ministerstwo Administracji i Cyfryzacji. Retrieved from https://isap.sejm.gov.pl/isap.nsf/download.xsp/WMP20130000121/O/M20130121.pdf (last accessed: 29 October 2021).

Polyák L (2015) Społeczny dostęp do przestrzeni publicznych Debata o dobrach wspólnych w Rzymie. *Autoportret*, 4(51), 18–21.

Popławska M (2013-05-24) Najlepsze te małe kina. *Res Publica Nowa*. Retrieved from: https://publica.pl/teksty/najlepsze-te-male-kina-37234.html

Popow M (2016) Organizacje pozarządowe, miasta i zmiana społeczna. *Studia Pedagogiczne*, 69, 173–187.

Public Space Charter (2009) Poznań: Third Congress of Polish Urban Planning, retrieved from: http://tup.org.pl/download/KartaPrzestrzeniPublicznej.pdf

Radzimski A (2014) Regionalne zróżnicowanie polityki mieszkaniowej: Przykład programu "Rodzina na swoim". *Rozwój Regionalny i Polityka Regionalna*, 28, 51–68.

Remuszko S (1990-07-03) Punkt krytyczny. *Gazeta Wyborcza*. Retrieved from https://classic.wyborcza.pl/archiwumGW/6015710/Punkt-krytyczny

Sadik-Khan J, Salomonow S (2017) *Walka o ulice: Jak odzyskać miasto dla ludzi*. Translated by Mincer W, Kraków: Wysoki Zamek.

Sadowy K (2014) Miasto jako dobro wspólne - perspektywa społeczna i ekonomiczna. *Myśl Ekonomiczna i Polityczna*, 4(47), 126–141.

Sagan I (2017) *Miasto: Nowa kwestia i nowa polityka*. Warszawa: Wydawnictwo Scholar.

Sapała M (2017-02-28) Polska masakra piłą mechaniczną. *Polityka*. Retrieved from https://www.polityka.pl/tygodnikpolityka/kraj/1695786,1,rzez-drzew.read

Sarzyński P (2015-05-12) Finaliści Nagrody Architektonicznej POLITYKI 2014. *Polityka*. Retrieved from: https://www.polityka.pl/tygodnikpolityka/kultura/1618709,1,finalisci-nagrody-architektonicznej-polityki-2014.read

Skwaradowska J (2016-03-19) Skludzej po swoim psie-akcja w Chorzowie. *Super Express*. Retrieved from https://www.se.pl/slask/skludzej-po-swoim-psie-akcja-w-chorzowie-wideo-aa-sctK-17wm-mbmb.html

Słodowa-Hełpa M (2015) Odkrywanie na nowo dobra wspólnego. *NierównoŚci Społeczne a Wzrost Gospodarczy*, 43(3), 7–24. http://doi.org/10.15584/nsawg.2015.3.1

Small Constitution of 1992 (1992) Journal of Laws no. 84, item 426.

Smarż J (2016) Pozwolenie na budowę, jako wyraz troski o dobro wspólne. In: Kułak-Krzysiak K, Parchomiuk J (eds), *Służąc dobru wspólnemu*. Lublin: Wydawnictwo KUL, 251–265.

Sobol A (2016a) The views of the young generation of a city as a common good. *Economic and Environmental Studies*, 16(4), 591–603.

Sobol A (2016b) Kategoria dobra wspólnego w zrównoważonym rozwoju miast. *Prace Naukowe Uniwersytetu Ekonomicznego we Wrocławiu*, 453, 87–95.

Sobol A (2017) Civic participation in terms of common goods. *Ekonomia i Środowisko*, 2(61), 8–18.

Socha R (2017-05-30) Jak działa gdańska rada ds. społecznych. *Polityka*. Retrieved fromhttps://www.polityka.pl/tygodnikpolityka/spoleczenstwo/1706420,1,jak-dziala-gdanska-rada-ds-spolecznych.read

Sokołowicz ME (2017) Miejskie dobra wspólne (commons) z perspektywy ekonomii miejskiej. *Studia Regionalne i Lokalne* 70, 23–40.

Solnit R (2016) *Hope in the Dark: Untold Stories, Wild Possibilities*. Chicago: Haymarket Books.

Sosnowski P (2017) Dobro wspólne jako przesłanka stanowienia prawa miejscowego. In: Zimmermann J (ed.), *Aksjologia prawa administracyjnego: Tom I*. Warszawa: Wolters Kluwer, 579–590.

Springer F (2016-06-11) Wolnoć, Tomku, we własnym domku. *Gazeta Wyborcza*. Retrieved from https://classic.wyborcza.pl/archiwumGW/8162453/Wolnoc--Tomku--we-wlasnym-domku

Stacul J (2018) The constructive power of slogans in post-socialist Poland. *Anthropologica*, 60(2), 494–506.

Stangret M (2008-12-23) Dobro wspólne? Nie istnieje. *Gazeta Wyborcza*. Retrieved from https://classic.wyborcza.pl/archiwumGW/5236189/Dobro-wspolne--Nie-istnieje

Staniszkis M (2009) Kolonizacja przestrzeni wspólnej. *Autoportret*, 4(29), 82–86.

Staniszkis M (2012) Continuity of change vs. change of continuity: A diagnosis and evaluation of Warsaw's urban transformation. In: Grubbauer M, Kusiak J (eds), *Chasing Warsaw: Socio-Material Dynamics of Urban Change since 1990*. Frankfurt: Campus Verlag, 81–108.

Stefanowski M (2004-08-12) Dlaczego tak brzydko. *Gazeta Wyborcza*. Retrieved from https://classic.wyborcza.pl/archiwumGW/4142165/Dlaczego-tak-brzydko

Surmiak-Domańska K (2007-10-01) Polsko, otwórz się: Interview with Maria Lewicka and Katarzyna Zaborska. *Gazeta Wyborcza*. Retrieved from http://www.archiwum.wyborcza.pl/Archiwum/2029040,0,4947722,20071001RP-DGW_D,Polsko_otworz_sie,.html

Sutowski M (2017) *Rok dobrej zmiany. Wywiady Michała Sutowskiego*. Warszawa: Wydawnictwo Krytyki Politycznej.

Szahaj A (2017-02-09) Gospodarka wolnorynkowa, czyli zapomniane dobro wspólne. *Rzeczpospolita*. Retrieved from https://www.rp.pl/plus-minus/art2928211-szahaj-gospodarka-wolnorynkowa-czyli-zapomniane-dobro-wspolne

Szczurek W (2011-03-03) Moją partią jest Gdynia! *Rzeczpospolita*. Retrieved from http://archiwum.rp.pl/1028250.html

Szułdrzyński M (2008-05-23) Cała władza w ręce lokalnych sitw. *Rzeczpospolita*. Retrieved from http://archiwum.rp.pl/778507.html

Szewczyk M (2018) Ład przestrzenny jako dobro wspólne. *Przedsiębiorczość i Zarządzanie*, 19(9/1), 91–102.

Szwed S (2012-06-23) Wspólnota. *Gazeta Wyborcza*. Retrieved from https://classic.wyborcza.pl/archiwumGW/7614444/Wspolnota

Szymanik G (2013-07-25) Marchewka w utopii. *Gazeta Wyborcza*. Retrieved from http://www.archiwum.wyborcza.pl/Archiwum/2029040,0,7767424,20130725RP-TDF,Marchewka_w_utopii,.html

Szyperska U (1999-01-16) Brzydota nie boli. *Polityka*. Retrieved from https://www.polityka.pl/archiwumpolityki/1825492,1,brzydota-nie-boli.read

Śpiewak J (2015-04-18) Leming naprawia polski kapitalizm. *Gazeta Wyborcza*. Retrieved from https://classic.wyborcza.pl/archiwumGW/8012264/Leming-naprawia-polski-kapitalizm

Środa M (2009-08-25) Skąd tylu obcych między nami? *Gazeta Wyborcza*. Retrieved fromhttps://classic.wyborcza.pl/archiwumGW/7072332/SKAD-TYLU-OBCYCH-MIEDZY-NAMI-

Trammer K (2019) *Ostre cięcie: Jak niszczono polską kolej*. Warszawa: Wydawnictwo Krytyki Politycznej.

Teister W (2017-02-23) Podatnicy zapłacą za drzewa sadzone na prywatnych działkach? *Gość Niedzielny*. Retrieved from https://www.gosc.pl/doc/3713207.Podatnicy-zaplaca-za-drzewa-sadzone-na-prywatnych-dzialkach

Thornberg R, Charmaz C (2013) Grounded Theory and Theoretical Coding. In: Flick U (ed.), *The SAGE Handbook of Qualitative Analysis*, 153–169.

Transcript of statements at the sessions of the second term of office of the Sejm of the Republic of Poland Wypowiedzi na posiedzeniach Sejmu RP II kadencji (1994) National Assembly on 22 September 1994, retrieved from http://orka2.sejm.gov.pl/Debata2.nsf/

Trincas M (2015) Laboratoria regeneracji miasta i terytorium. *Autoportret*, 4(51), 29–35.

Twardoch A (2019) *System do mieszkania*. Warszawa: Fundacja Nowej Kultury Bęc Zmiana.

Wajrak A (2017-06-24) Kto odpowie za trzy miliony ściętych drzew. *Gazeta Wyborcza*. Retrieved from https://classic.wyborcza.pl/archiwumGW/8270104/KTO-ODPOWIE-ZA-TRZY-MILIONY-SCIETYCH-DRZEW

Wielgo M (1999-01-20) Tylko bez samowoli. *Gazeta Wyborcza*. Retrieved from: https://classic.wyborcza.pl/archiwumGW/644410/Tylko-bez-samowoli

Wielgo M (2013-02-26) Ogrody do obrony. *Gazeta Wyborcza*. Retrieved from http://www.archiwum.wyborcza.pl/Archiwum/2029040,0,7710259,20130226RP-DGW,Ogrody_do_obrony,.html

Wielgo M (2015-07-28) Budowa mieszkań tylko według gminnego planu. *Gazeta Wyborcza*. Retrieved from: http://www.archiwum.wyborcza.pl/Archiwum/2029040,0,8050047,20150728RP-DGW,Budowa_mieszkan_tylko_wedlug_gminnego_planu,.html

Wiśniak M (2003) Zabieg: obieg. *Autoportret*, 3(4), 22–25.
Wiszniowski J (2019) Krajobraz jako dobro wspólne. In: Drapella-Hermansdorfer A, Mycak O, Surma M (eds), *Krajobraz jako wyraz idei i wartości*. Wrocław: Oficyna Wydawnicza Politechniki Wrocławskiej.
Witosz B (2016) Czy potrzebne nam typologie dyskursu? In: Witosz B, Sujkowska-Sobisz K, Ficek E (eds), *Dyskurs i jego odmiany*. Katowice: Wydawnictwo Uniwersytetu Śląskiego, 22–30.
Wodecka D (2017-04-29) Wściekli i solidarni. Interview with Robert Biedroń. *Gazeta Wyborcza*. Retrieved from https://classic.wyborcza.pl/archiwumGW/8257000/WSCIEKLI-I-SOLIDARNI
Wojnicki J (2020) Tendencje centralizacyjne w Polsce po 1990 roku – ciągłość i zmiana. *Polityka i Społeczeństwo*, 2, 26–37.
Wrabec P, Socha R (2006-03-25) Wspólna nie dla wszystkich. *Polityka*. Retrieved from https://www.polityka.pl/archiwumpolityki/1810230,1,wspolna-nie-dla-wszystkich.read
Wybieralski M (2013a-01-19) Ciałoblokada: Interview with Joanna Erbel. *Gazeta Wyborcza*. Retrieved from www.archiwum.wyborcza.pl/Archiwum/2029040,0,7693703,20130119RP-TWY,Cialoblokada,wywiad.html
Wybieralski M (2013b-11-29) Między tandetą a granitem. *Gazeta Wyborcza*. Retrieved from http://www.archiwum.wyborcza.pl/Archiwum/2029040,0,7818751,20131129RP-DGW,MIEDZY_TANDETA_A_GRANITEM,.html
Zaiček M (2015) Pieśń dla Machnáča. *Autoportret*, 4(51), 82–85.
Zarycki T (ed.) (2013) *Polska Wschodnia i orientalizm*. Warszawa: Wydawnictwo Naukowe Scholar.
Zieliński MA (2015-05-02) Polska budzi się 40 lat za późno. Interview with Jan Sowa. *Gazeta Wyborcza*, retrieved from: http://www.archiwum.wyborcza.pl/Archiwum/2029040,0,8017244,20150502RP-DGW,Polska_budzi_sie_40_lat_za_pozno,.html
Ziemacki J (2018-10-05) Nie tylko dla koneserów zabytków. *Rzeczpospolita*. Retrieved from https://www.rp.pl/nieruchomosci/art9609721-nie-tylko-dla-koneserow-zabytkow
Zinserling I, Grzelak J (2008) Orientacje społeczne i orientacje kontroli a wrażliwość Polaków na dobro wspólne. In: Rutkowska D, Szuster A (eds), *O różnych obliczach altruizmu*. Warszawa: Wydawnictwo Naukowe Scholar, 110–133.
Żakowski J (2010-06-05) *Wielki odpływ*. *Polityka*. Retrieved from https://www.polityka.pl/archiwumpolityki/1920472,1,wielki-odplyw.read
Žižek S (2009) *First As Tragedy, Then As Farce*. London: Verso Books.
Żurawik A (2014) “Public interest” and “common good”: general clauses in both Polish and European Union law. *Polish Review of International and European Law*, 3(1–2), 71–97.

4 'Going back to the obvious?' Interpretations of common good by urban actors in Gdańsk, Kraków, and Łódź

Case study cities: selection and overview

According to what has been established in previous chapters, the paradigm of the city as a commons has been gaining prominence in Polish public discourse for over a decade now. In practical terms, it means that it has been increasingly talked, written, and read about. However, while the results of discourse analysis set a comprehensive background on how it emerged and developed, subjective perspectives of various actors involved in conceptualisation and production of the urban common good remain less known. To get a more detailed picture of how urban stakeholders interpret and exercise this notion, it is thus necessary to switch to the local scale and concrete urban issues from concrete cities.

For the purposes of this study, I have decided on a triad of major Polish cities—Gdańsk, Kraków, and Łódź. The adopted multiple case-study approach pursues several research objectives. Firstly, a closer look into the three urban microcosms gives a broader insight into the development of public discourse on common good as understood locally. Secondly, even if generalising from a single case is justified in social sciences (Flyvbjerg 2006), a comparative perspective allows one to observe common patterns but also capture variations. Lastly, another advantage of a cross-case analysis is its applicative value as it enables learning from multiple cases, and gathering critical evidence necessary for policy change (Khan and Van Wynsberghe 2008).

The selection of Gdańsk, Kraków, and Łódź as case studies was guided by numerous reasons. First of all, as second-order cities and regional metropolies of similar size (Figure 4.1), they make up informative urban laboratories of social and spatial transformations.[1] At the same time, these processes of change are relatively less researched than in the capital city of Warsaw, whose development is governed by its own rules, and therefore not always fit for yardstick purposes. Although the three urban centres share several common features, they differ significantly in many respects. Due to their distinct origins and diverse paths of development and external factors, each of the three cities carries a considerably unique historical and cultural baggage. The mediaeval origin of Gdańsk and Kraków rendered their growth patterns

DOI: 10.4324/9781003089766-7

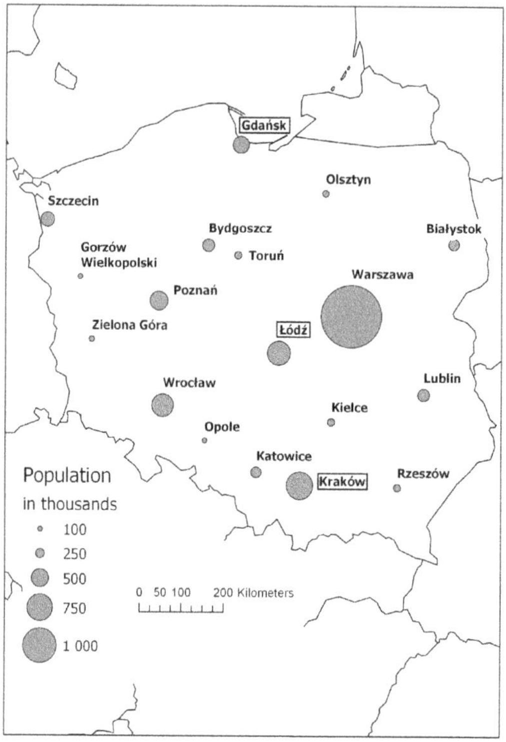

Figure 4.1 Location of the case-study cities within regional capitals in Poland.

quite different from those of Łódź—a late bloomer, whose rapid expansion in the 19th century was driven by modern industrial capitalism. In addition, following the three consecutive partitions of Poland in the late 1700s, each of the cities was claimed, and then-on shaped, by a separate superpower: the Kingdom of Prussia (Gdańsk), Austria (Kraków), and the Russian Empire (Łódź).[2] The legacy of the partitions still determines regional socio-economic disparities across the country (Gorzelak 2006). For instance, according to a contemporary study of local citizen engagement (Peisert 2013), Poles inhabiting the lands formerly annexed by Austria are far more active than their compatriots from the former Russian Partition.

More importantly, in the context of the focus of this book, substantial differences between the case studies follow from the cities' diverse trajectories after 1945 and 1989. Throughout the second half of the 20th century, **Gdańsk** was an arena of pivotal transformations in matters of society, culture, space, and politics. After World War II (WWII), the severely destroyed and depopulated city was rebuilt and (re)appropriated by newcomers—mostly from central Poland and Eastern Borderlands—who struggled to settle in their new home (Perkowski 2013). The next two decades brought rapid industrialisation and urbanisation, the pressures of which fuelled the construction of large, prefabricated housing estates, and had a profound impact on the urban landscape, while also improving the quality of urban life. The resulting overall spatial development consolidated Gdańsk, Sopot and Gdynia into Tricity, the core of a fast-growing agglomeration. Finally, the workers' strikes of the 1970s and 1980s activated the centripetal social dynamics towards common good, galvanising thousands of regular citizens to take to the streets and protest the ruling party, which largely contributed to the fall of communism in Poland.

The turn of the century saw only isolated cases of civic engagement for the sake of collective interests. Quite symbolically, the site occupied by the eponymous Gdańsk Shipyard—the foremost arena of the Solidarity movement—was privatised and left largely out of public control. Until around the mid-2010s, the main priorities of the local urban policy were set on large-scale investment in transport infrastructure, especially with regard to the 2012 UEFA European Football Championship co-hosted with Ukraine (Karwacka 2010, Kostrzewska and Rembeza 2012). Housing-wise, the city authorities favoured greenfield development in the peripheral districts, which fuelled fragmentation of space and a chaotic expansion of gated housing estates (Polanska 2010). At the same time, Gdańsk was the first large city in Poland to experiment with innovative tools of governance—participatory budgeting and citizens' assemblies (Glejt-Uziębło and Uziębło 2018).

Unlike Gdańsk, **Kraków** suffered relatively less war damage and population exchange after WWII. However, in 1951, the communist regime's efforts to rebrand the city's traditional image led to expansion through the incorporation of Nowa Huta—a model socialist industrial town developed from a rural suburb to accommodate the staff of the flagship Lenin Steelworks (see Chapter 2). Yet, while Kraków's social composition shifted to include more blue-collar workers, the petite bourgeoisie and intelligentsia managed to retain their position in the city, which continued to be perceived as one of the key cultural and scientific centres in Poland (Klich-Kluczewska 2005).

Within the case-study triad, Kraków has been the least affected by the costs of the systemic transformation in terms of demographic and economic trends. The population dynamics remained roughly stable, and the local economy has increasingly developed by the expanding tourism sector (Tracz and Semczuk 2018). Culture-led regeneration further enhanced this orientation and strengthened the city's internationally competitive position (Murzyn 2006). However, Kraków's return to self-governance was particularly strenuous, and the neoliberal agenda all the more embraced (Purchla 2018). In effect, the mainstay of the Polish bourgeoisie also became a breeding ground for a multitude of urban movements, with their raucous advocacy for the right to housing, clean air, and against mega-events planned by the local authorities. For instance, the city of Kraków was forced to withdraw its application as a candidate host city for the 2022 Winter Olympics when a bottom-up initiative, Kraków Przeciw Igrzyskom (Kraków Against the Olympics), had culminated in a local referendum in which an almost 70-per cent majority voted 'no' to hosting the Games in the city (Mazurkiewicz 2021).

Finally, there is **Łódź**, the former textile 'promised land' or 'Polish Manchester' (cf. Liszewski and Young 1997). Considerably depopulated during WWII, the city's urban infrastructure, including the industrial facilities, was relatively undamaged—especially in comparison with the almost completely ruined Warsaw. As such, it was briefly considered as a potential new seat of the Polish government. Under socialist economic planning, Łódź quickly reclaimed its pre-war standing of a major industrial city and flourished for several more decades. The existing capitalist infrastructure was symbolically handed over to the people by means of nationalisation—luxurious residences were used for kindergartens and public offices, private gardens, and parks were made widely accessible, and the industrial plants were no longer committed to simple profit-making. New urban developments emerged on the city fringe, and the compact inner city, with its housing stock still inhabited but sentenced to decay through poor maintenance and underinvestment, became fragmented by wide traffic arteries (Dankowska 2016). Since the late 1940s, Łódź had consistently developed its film-making industry. However, regardless of the fact that by the end of the 1980s the city was the second-largest urban centre in Poland after Warsaw, with the number of inhabitants exceeding 850,000, its monofunctional reliance on the textile industry translated into an acute economic degradation post-1989.

The eventual decline of the local economy during the transition period triggered structural unemployment and profound depopulation, marking the 1990s in Łódź with an especially severe and multidimensional urban crisis (Jażdżewska 2010). Expenditure on the maintenance of council housing stock, old and underinvested, only added to the city's financial woes, and to a large extent hampered further urban development. The local renewal programmes adopted at the beginning of the 2000s were to address this issue but they mostly aimed at improving the built environment, without much consideration for the social aspects of regeneration (Lamprecht 2016). Particular emphasis was put on pre-war tenement houses renovation, and ramping up the aesthetic charms of Piotrkowska Street, the centrally located high street. Recent large-scale investment in infrastructure projects, and the construction of a central business district under the name New Centre of Łódź, enabled by public–private partnerships and EU funding, have been recognised as efforts towards changing the city's image to make it 'aligned with the demands of the day' (Śmiechowski and Burski 2018: 256). The private redevelopment of a former textile factory into Manufaktura, a commercial complex, has become a symbol of the city's post-socialist renewal (Tobiasz-Lis 2016).

As each of the case-study cities' social, economic, and cultural backgrounds are unlike the other, the systemic transition has affected them to varying degrees, which is reflected in their current socio-economic characteristics as presented in Table 4.1. It is Łódź which comes out as the most deprived within the triad, owing it to the largest depopulation since the mid-1990s, the most rapidly ageing population, the highest unemployment rate,

Table 4.1 Comparison of basic socio-economic characteristics in the three case-study cities

Chosen indicators	*Year(s)*	*Gdańsk*	*Kraków*	*Łódź*
Area (sq. km)	2019	262	327	293
Population (000)	2019	470.9*	779.1	679.9
Population change (%)	1990–2019	1.2	3.8	−19.8
	2005–2019	2.8	2.8	−11.4
	1990–2004	−1.3	1.0	−8.8
Age structure (%)				
0–17		17.8	17.2	15.1
18–59/64	2019	58.0	59.1	56.4
60/65+		24.2	23.7	28.6
Unemployment rate (%)	2019	2.4	2.0	4.8
	2004	11.5	7.5	18.4
Total income/expenses per inhabitant (000 PLN)	2019	7.7/8.0	7.6/8.0	6.6/6.9
	2002	2.0/2.2	1.9/2.1	1.9/2.1

* Tricity: 751.1 (incl. Gdynia: 246.3 and Sopot: 35.7).

and the lowest income per inhabitant as of 2019. As far as these indicators are concerned, Gdańsk and Kraków are much better off, and mostly on par, except for slightly different population dynamics. Yet, as much as their biographies and current socio-economic situations vary, all the three cities have been struggling with quite similar challenges which plague post-socialist cities. Unchecked urban sprawl is one of such headaches, aging population is another. Moreover, all the cities had to tackle high unemployment as they endured an ever-growing disparity between incomes and increasing municipal expenditure.

In all the three case-study cities urban policies have been likewise dominated by the neoliberal philosophy. As already signalled, they have all seen prioritisation of mega-events validate large-scale investment, and charter course for spatial and infrastructural development. For instance, high hopes associated with the eventually failed Expo bid had translated into a vision for the New Centre of Łódź. In Kraków, out of 16 pieces of investment related to the Euro 2012, only three brought upgrades to existing sports facilities, while almost a half concerned the expansion of road infrastructure for private-car-based transport (Jarzębiński 2014). In Gdańsk, the approaching football event spurred a hasty revitalisation of a dilapidated neighbourhood adjacent to the championship venue (Kamrowska-Załuska and Kostrzewska 2014).

The increasing civic engagement and activity of urban movements in Gdańsk, Kraków, and Łódź (Grabkowska and Makowska 2011, Kubicki 2011, Madejska 2014) applied bottom-up pressure on the city as a commodity paradigm. Partly as a result, around the mid-2010s the initially overlooked notion of common good made its way to the local legal and administrative discourses. Its unfolding over the research period may be traced in the evolution of such documents as development strategies and studies of the conditions and directions of spatial development and urban regeneration programmes. The 2004 version of the Development Strategy adopted in Gdańsk situates concern for common good in a very general context, mentioning it as one of the factors that boost social capital, which in turn translates into socio-economic development (Strategia Rozwoju Gdańska...2004: 49). Conversely, according to one of the operational aims of the 2014 Strategy, which underlines the significance of continuous and permanent consultations with citizens, Gdańsk aspires to become 'a place where common interests of all stakeholders are agreed upon in a dialogue' (Gdańsk 2030 Plus... 2014: 93). Similarly, the Kraków 2030 Development Strategy conjures up the image of the city as a common good interpreted as a strong local government, a community of residents highly involved in local governance, driven by civic solidarity, and cultivating shared values to achieve even the most difficult common goals (Kraków Development Strategy 2018: 118). Correspondingly, the concept of the 'city as a common good' was adopted in the Strategy of Integrated Development of Łódź 2020+ (2012) as one of the three strategic aims, directed at building high-quality social capital, and developing citizen participation in decision-making. The means to this end include, inter alia, creating effective channels of communication with citizens, promoting self-organisation of social actors, providing citizenship education for the young, supporting systemic activation of females and the elderly, and introducing participatory budgeting as a tool for local co-governance. This goes in line with one of the operational goals featured in a more recent version of the local revitalisation programme—'community-building around the common good' (Program Rewitalizacji Łodzi... 2016: 96)—which is to be achieved by local urban regeneration initiatives serving the immediate surroundings or wider communities.

Even if local regulations express only declarative concern for common good, the signs of urban priority reorientation are already visible. In some areas, they are more distinct than in others, but in general, some of the recent socio-economic indicators reflect an ongoing change (Table 4.2).

Possibly the worst situation concerns managing the municipal **housing** stock, the share of which in all the apartments dwindled over the research period in each of the three cities, while affordable housing programmes were generally missing from the local policy agendas. Although a comparison of the quantity of new municipal flats in 2000 and 2019 suggests a reversal of the trend, in other years the numbers went up and down.

As far as **public transport** goes, the era of large-scale investment in individual car transport infrastructure seems to be waning. The latest flagship developments are now rather dedicated to multi-modal public transport, as in the case of the tram station dubbed Unicorn Stable due to its flamboyant design (Figure 4.2), and a new railway tunnel which will serve the local 'metro' in Łódź (Szpakowska-Loranc and Matusik 2020). Accordingly, a comparison

Table 4.2 Symptoms of the city as a commons paradigm settling in the three case-study cities

Chosen indicators	*Year(s)*	*Gdańsk*	*Kraków*	*Łódź*
Share of council flats in total number of flats (%)	2018	7.6	3.8	12.3
	1995	30.8	17.4	28.6
New council flats	2019	69	179	16
	2000	0	0	15
Bus lanes (km)	2019	5.4	30.9	24.2
	2013	2.0	23.7	10.7
Cycle lanes per 100 sq. km (km)	2019	74.8	72.2	56.6
	2011	34.4	33.1	24.8
Share of urban green areas in total area (%)	2019	3.4	5.8	5.7
	2004	2.1	3.8	6.1
Planted trees	2019	540	2937	1656
	2004	82	2040	415
Share of total area covered by zoning plans (%)	2019	65.8	65.0	24.4
	2009	61.3	16.7	4.6
Issued decisions on building conditions	2019	385	669	1452
	2009	388	2203	2161

Figure 4.2 Attractiveness of public transport revisited: Unicorn Stable in the city centre of Łodź.

of the total length of bus lanes shows an upward tendency between 2013 and 2019 in all three cities, even though Kraków and Łódź are much more advanced in this regard than Gdańsk. During a similar period, the length of cycle lanes per 100 square kilometres more than doubled in each city. Likewise, the popularity of pedestrianised zones in city centres has been on the rise.

The amount of **green infrastructure** fluctuated in Łódź but increased rather steadily in the other two cities. The distribution of the numbers of trees planted each year between 2004 and 2019 shows irregular patterns but an overall positive trend. In 2015, bottom-up initiatives for the preservation of open green spaces in Kraków propelled the local authorities to establishing a new organisational unit to manage urban greenery.

Public space indicators are not provided by the Polish statistical office database but some approximation is provided by the data on the share of areas covered by zoning plans, which has been on the rise. Unsurprisingly, in Łódź, where the coverage rate was the lowest among the three cities, the number of issued decisions on building conditions and land development was the highest. Yet, none of the three cities assured wide participation of citizens in the planning processes for recent municipal or public–private investment in public space. The only effective way for citizen actors to have an impact locally is through participatory budgeting (Rzeńca and Sobola 2018, Szczepańska et al. 2021), however, with no guarantee of a democratic effect (see Chapter 2).

The regeneration programmes in all the three cities have neglected the social dimension in favour of aesthetics and economics, not seldom siding with gentrification (Gądecki 2012, Jakóbczyk-Gryszkiewicz 2015, Marcińczak et al. 2012, Murzyn-Kupisz 2012). Some reversal of this tendency has been lately propelled by the pro-participatory regulations of the Act on Urban Regeneration, and citizen engagement in participatory budgeting. Again, using the example of Łódź, out of over a hundred currently implemented regeneration tasks, around a half promise to accomplish a goal defined as 'building a community of residents cherishing the common good' (Program Rewitalizacji Łodzi… 2018: 99). These mostly public initiatives include, for example, development of the social rental housing sector, creating inclusive and user-friendly public spaces, reusing facilities of EC-1, a former power-plant, for cultural and educational purposes, or finding new functions for old community centres (*domy kultury*).

As of yet, **spatial justice** has not been addressed holistically in any of the case-study cities. However, in all three of them, some action has been taken to that end. For instance, Gdańsk was the first city to implement two inclusionary measures empowering marginalised citizens—the Immigrant Integration Model (2016) and the Model for Equal Treatment (2018). Moreover, all three cities developed urban climate change adaptation plans as part of a project held under the auspices of the Ministry of the Environment. The said documents have been accepted by the local councils, and the measures thereof are being currently implemented.

Interviewing urban stakeholders in Gdańsk, Kraków, and Łódź

As evidenced so far in this book, urban arenas are occupied by actors who hold dissimilar power positions, play diverse roles, and pursue unlike interests. Hence, different users of the city tend to understand and practice the idea of urban commonality differently. Although a few existing studies from Western European cities establish that groups as decision-makers or planners employ the notion of common good according to their own grasps and needs (Häikiö 2007, Murphy and Fox-Rogers 2015), no attempts at investigation of several standpoints have been made so far. The research conducted for the purposes of this chapter aims to fill this lacuna.

The adopted purely qualitative approach consists of semi-structured individual interviews with representatives of five groups of urban stakeholders. As a research method, interviewing allows one to explore causal relations based on subjective human experiences and attitudes which are otherwise hard to reach (Peräkylä 2009: 325). In the presented study, these concerned individual understandings of the urban common good, rationales, values, and the preferences shaping them, and the perceived ways of how the common good is practised and achieved (or not). The intention of unpacking the interviewees' conceptualisations of urban common good was to understand different logics behind them. To avoid operating with only abstract ideas, in each of the three cities a portion of questions concerned specific issues identified as potentially related to urban common good, in line with the assumption of the concept of concrete narrative of Mergler et al. (2013).

Since the focus of this book is on citizens as key urban actors, the selection of stakeholder groups was guided by their perceptions of who else populates the urban microcosm, and to what extent. Such stance was adopted in a study of the composition of urban agoras in Gdańsk, Gliwice and Wrocław (Kajdanek 2017: 173–174). Although the results vary slightly between the cities, the respondents generally agree on the following typology of participants in the local debates concerning the city: local authorities (81.8 per cent of indications), citizens (74.6 per cent), the media/journalists (55.1 per cent), investors/developers (46.8 per cent), architects, planners, and scholars (43.9 per cent), NGOs (40.4 per cent), urban movements (39.0 per cent), and cultural circles (33.9 per cent).

As the media and academic perspectives had already been covered in detail in the discourse analysis presented and discussed in Chapter 3, my decision was to dispense of the media/journalists category and to narrow down the semantics of 'experts' to architects and planners employed in municipal institutions. Moreover, due to their resemblance, the categories of NGOs and urban movements were consolidated into one. Accordingly, five groups of major local stakeholders were recognised for the purposes of field research: citizens, urban activists, decision-makers, municipal planners, and local entrepreneurs. Such a collection of perspectives allows to observe differences between the groups but also makes it possible to compare and contrast two types of discourse—constructed and communicated from the top down (decision-makers and planners), and from the bottom up (citizens, urban activists, and entrepreneurs).

According to Lia Karsten (2009: 318), such a distinction may be crucial for understanding incongruities between the two perspectives, potentially contributing to their perception as conflicting and irreconcilable.

As already indicated, the interviews were semi-structured, which means they involved only a few open-ended questions, inciting a rather unconstrained dialogue, and leaving the interviewees much space for self-reflection. In general, the interview protocol was organised into two thematic sections. The first one revolved around such topics as the notion of common good in a city, both in theory and practice; the conditions which reinforce and undermine urban common good; the qualification of urban space as commons, private and no one's; and the spatial conflicts resulting from these distinctions. The second section focused on a more specific discussion of selected issues relating to urban common good. In order to identify them in each of the cities, I resorted to a basic media content analysis. The local editions of *Gazeta Wyborcza* in Gdańsk, Kraków, and Łódź were searched for articles pertaining to concrete and relevant local problems debated in this context throughout the preceding decade.

As preparation for the field phase of the case-study research began in 2016, the media content analysis covered the years 2006–2015, placing the 2011 caesura at the centre. The collection of the sample followed the same procedure as the one outlined in Chapter 3—it was based on searches by the same keywords related to urban common good. The number of articles containing the keywords, showing a general upward trend in all the local editions, amounted to 16 in *Gazeta Wyborcza Trójmiasto*, 18 in *Gazeta Wyborcza Kraków*, and 8 in *Gazeta Wyborcza Łódź*. The most frequent references informed a catalogue of specific themes identified to be prominent in the print media discourse on the local scale, out of which themes three per each case-study city were selected for further investigation.

Gdańsk: post-shipyard areas, city-centre courtyards, and residential gating

The articles in *Gazeta Wyborcza Trójmiasto* most often featured common good in the context of progressing processes of metropolisation within the Tricity region and concerned the shared interests of Gdańsk, Sopot, Gdynia, and the surrounding gminas. However, because this phenomenon reaches beyond the local scale, it was left out from the analysis. Other distinct thematic threads mostly employ the notion of urban common good to spatial management. They include the debate on the redevelopment of the former shipyard site, the management and use of the courtyards in the city centre, and conflicts over residential gating.

The **post-shipyard areas** cover around 70 hectares of brownfields which emerged after the 1990s scale-back of the Gdańsk Shipyard. They are located close to the historical city centre and include a long stretch of waterfront. Both their attractiveness and perception as common heritage are very much rooted in the Shipyard's symbolic significance as the birthplace of the Solidarity movement and a site of resistance against the communist government (Lorens and Bugalski 2021). Today, they are an important element of the image of the city, for citizens and tourists alike. Formerly owned by the state, the land was privatised in 1999, and intended for redevelopment into a mixed-use neighbourhood under the name Młode Miasto (the Young Town). However, so far new investment has proceeded slowly—the flagship European Solidarity Centre was completed in 2014, and the transit road of Popiełuszki Street, criticised for prioritisation of car traffic at the cost of pedestrians, for long have been the sole major developments in the area. As of today, the first housing estate on the former shipyard premises is under construction. In the meantime, demolitions of historical buildings, and the municipality's loss of control over the redevelopment process gave rise to public outcry. One of the charges against the local authorities concerns the fact that public debate over the zoning plan, adopted in 2001, was practically non-existent. The plan meant adoption of regulations which are assessed today as misguided and out-of-date (Krzymiński 2015). Another aspect of the perceived commonality of the post-shipyard areas consists in their ongoing cultural colonisation as of the early 2000s and ranging from more and less informal artists' residencies, through location of a new museum of modern art, to concert venues and clubs. The site is thus subject to multiple tensions between the neoliberal vision underpinned by the principle of profit maximisation, and the more recent attempts at recognising and protecting the common heritage via organic, bottom-up initiatives (Szmytkowska and Nowicka 2015, Prawelska-Skrzypek and Morgan 2020) (Figure 4.3).

Figure 4.3 Recommoning of the former Gdańsk Shipyard? Developer's invitation to sightseeing.

The controversy over **the city-centre courtyards** relates to the public–private division of urban space in the Main Town of Gdańsk, where interests of different users clash on a daily basis. It has its origins in the socialist planning principles which guided the post-war reconstruction of the historical city centre. To provide access to sunlight and open green space, instead of a full reconstruction of the original townhouses, only three out of the four original perimeters were restored. The resulting semi-private spaces within the blocks were to be used by residents but managed by the municipality. Over the years, with maintenance leaving a lot to be desired, the condition of these common yards was progressively deteriorating (Szechlicka et al. 2016). Moreover, the reintroduction of commercial uses of space to the city centre following 1989 sparked conflicts between residents, local entrepreneurs, and, occasionally, tourists. To address the increasing discontent of the former group, in 2018, the local authorities launched an incentive programme aimed at promoting civic responsibility over the courtyards (*Gdańskie Nieruchomości* 2022) and offering funding for renovation and fencing of the yards to the homeowner associations wishing to lease the courtyards. This in turn raised the debate over the right of public access to the formerly open space.

The problem of courtyards in the Main Town is connected to the more general processes of **residential gating** in Gdańsk. Originally native to the greenfield development of peripheral districts, the phenomenon was noticed and discussed, mostly by scholars, in the context of residential segregation and fragmentation of urban space (Polanska 2010). The downsides of common space enclosures became more apparent to residents once secondary gating trends, affecting the existing residential estates, developed in the inner city, and were espoused by large-scale housing estates (Grabkowska and Szmytkowska 2021). While the Landscape Code, adopted in 2018 (Uchwała Krajobrazowa Gdańska 2018), regulates the problem of fencing off property to some extent, especially in terms of aesthetics, the issue has remained a bone of contention in the local public sphere. More recently, the local developers tend to walk away from this trend, which seems to confirm its decreasing popularity among local homebuyers.

Kraków: air quality, urban green areas, and touristification

Reference to urban common good in the articles published in *Gazeta Wyborcza Kraków* mostly touched upon environmental resources—namely air pollution, and diminishing urban green area—as well as the issues radiated by touristification processes in the inner city.

The problem of **air quality** is anything but new in Kraków. Due to the city's specific geographic location, air pollution, generated by extensively used low-quality feedstock for household heating boilers with car traffic earning an honorary mention in the category, reaches the highest levels of all Polish cities (Kuchcik and Milewski 2018). As the local authorities had done little to address this problem, a grassroots anti-smog initiative, Krakowski Alarm Smogowy (Kraków Smog Alert), was established in 2012 to campaign for a blanket ban on household use of coal in the city. Voicing public concern for the quality of urban air, and demanding systemic solutions to the problem, through social-media campaigns, and anti-smog rallies, the activists also raised awareness of the air being a common urban resource. Eventually, in 2018, the campaign proved successful; however, as in other Polish cities (Frankowski 2020), questions of its exclusionary effects related to energy poverty remain valid.

As evidenced in the statistical data (Table 4.2), the share of the total area of **urban greenery** in Kraków in 2019 was the highest among the case-study cities. Nevertheless, a considerable proportion of the area, owned by the Catholic Church and enclosed, remains inaccessible to residents. Moreover, many of the existing open green spaces are being grabbed by developers (Hrehorowicz-Gaber 2013). One of the most notorious cases concerns the recreational grounds surrounding the water reservoir in the shut-down quarry of Zakrzówek (Figure 4.4). Plans for the development of a private residential estate were actively protested against, and abandoned as a result. One of the protesting groups, named the Modraszek Collective after a protected species of butterfly found in the area, was made up of local activists and citizens, who would appear in public spaces sporting blue cardboard wings (Janik 2020). Currently, the area is being transformed into a municipal park, however, still without significant public participation.

The last theme covers a spectrum of urban transformations grouped under the umbrella term of **touristification**. As mentioned earlier, tourism has recently skyrocketed in Kraków—only between 2010 and 2015, the annual number of visitors to the city increased from almost 8.2 to 10.1 million (Grabiński 2015). Today, mass tourism is as much an asset as it is a problem, generating much income but also bringing about the commercialisation of urban space and gentrification (Kruczek 2018). In response to the former, as early as in 2012, the historical city centre became legally protected as a culture park, which attached multiple restrictions to the commercial use of the existing buildings and their surroundings, particularly with regard to their

(a) (b)

Figure 4.4 Zakrzówek in Kraków: site of recreation in 2010 (a) and protest in 2011 (b).

aesthetics (Uchwała w sprawie… 2010). The latter issue has thus far remained unaddressed by the local authorities.

Łódź: municipal tenements, New Centre of Łódź, woonerfs

The discourse on urban common good as recorded on the local pages of *Gazeta Wyborcza* in Łódź was dominated by issues broadly connected to urban regeneration—the renovation of municipal tenements, the development of the New Centre of Łódź, and embracing the concept of woonerf as a strategy for neighbourhood revitalisation.

After 1989, the commonality of **municipal tenements** in Łódź first resurfaced as an issue of aesthetics to only later acquire a new meaning in the context of their value as common housing resources. The key objective of the City of (100) Tenements programme (*Mia100 Kamienic* 2011), namely a drive for renovation of municipally owned tenements of high cultural value, resulted in considerable displacement of the less affluent tenants from the city centre, rendering the municipality of Łódź the most effective gentrifier among Polish cities (Jakóbczyk-Gryszkiewicz 2015). Although more recently the local authorities have abandoned this strategy, the public debate over the right to municipal housing in the most attractive urban localisations is far from over.

The megaproject of a central business district under the name **New Centre of Łódź** (*Nowe Centrum Łodzi*) evolved from an idea put forward by a local foundation, that is, to reuse a centrally located former power plant as a cultural hub in celebration of the city's film-making heritage. In 2008, the 'EC1 Łódź—the City of Culture' plan was voted in by the city council, later expanding to include a thorough refurbishment of the nearby Łódź Fabryczna railway station—which has been lately buried underground and morphed into a multimodal transport hub—and erection of an office complex Brama Miasta (Gate to the City). According to the local authorities, the currently implemented large-scale development addresses the lack of a traditional city centre in Łodź—to date substituted for by Piotrkowska, the high street (Brzeziński 2011). Yet, the inclusionary aspect of the whole undertaking has been decried by local experts and citizens alike.

Against this backdrop, the transformation of one of the inner-city streets into a **woonerf** is all the more notable. Implemented as a bottom-up initiative mediated via participatory budgeting in 2014, its unforeseen success led to a proliferation of undertakings of this type (Zdyb 2017). It also prompted the local authorities to include a combination of the woonerf formula, and participatory approach in the most recent version of the regeneration programme as a tool for 'quality improvement and creation of public and semi-public (parapublic) spaces arranged with input from the citizens' (Program Rewitalizacji Łodzi 2016: 106–107).

Whereas the outlined problems are city-specific, at the same time they relate to the six areas of urban commonality identified in Chapter 2. All of them fit into more than one frame of reference, although to varying degrees (Table 4.3). What further links them is the mediation of urban common good

Table 4.3 Position of the analysed issues of urban common good within the six thematic frames of reference

	Gdańsk			*Kraków*			*Łódź*		
	Post-shipyard areas	*Courtyards in the city centre*	*Residential gating*	*Air quality*	*Urban green areas*	*Touristification*	*Municipal tenements*	*New Centre of Łódź*	*Woonerfs*
Housing			x			x	xxx		
Public transport	x			xx				x	xx
Green infrastructure		x		x	xxx				x
Public space	xx	xxx	xxx		x	xxx	x	xxx	xxx
Urban regeneration	xxx					x	xxx	xxx	xx
Spatial justice	x	xx	xx	xxx	xx	xxx	xxx	xxx	x

xxx – high relevance, xx – medium relevance, x – low relevance.

in the sense of access to and management of a variety of urban resources—unpolluted air, green areas, municipal dwellings, and space in general. The examples of the post-shipyard areas in Gdańsk or the greenery in Cracow show that the perception of commonality is not necessarily connected to the actual ownership status, nor with the economic categories of (non-)excludability and (non-)rivalrousness.

Altogether, 44 individual semi-structured interviews were carried out between April and August 2017. They proportionately covered the three cities and the five groups of urban stakeholders. Fifteen each were conducted in Gdańsk and Kraków, while fourteen took place in Łódź. The original plan for approaching three representatives of each stakeholder category in every city was verified by the respondents' availability. In the end, the sample counted ten planners (4 from Łódź, 3 from Gdańsk, 3 from Kraków), nine activists (3 per city), nine decision-makers (3 per city), nine citizens (3 per city), and eight entrepreneurs (4 from Kraków, 3 from Gdansk, 1 from Łódź).[3]

The interviewees were selected by purposive sampling, which relies on subjective choice. Urban activists, planners, decision-makers, and entrepreneurs, based in Gdańsk, Kraków, and Łódź, were approached based on their specialism, and impact on urban space. The interviewed **activists** included local NGOs founders and members, socially engaged artists, as well as employees of cultural institutions involved in socio-cultural animation and interventions in urban space. The **planners** were represented by various municipal institutions employees—bureaus of urban development, departments of architecture and urban regeneration, and units responsible for urban infrastructure. The category of **decision-makers** encompassed city and district councillors and a deputy mayor responsible for spatial policy. The owners of pubs, cafés, and clubs, local real estate developers, and founders of social cooperatives formed the group of **entrepreneurs**.

The choice of the interviewed **citizens** was more random, depending on the availability of contacts. In each city, they were selected from three generational groups (25–39 years, 40–59 years, and 60+ years). Such a criterion was adopted for two reasons. Firstly, people of different ages tend to follow different lifestyles and perceive urban space differently. Secondly, their age determines the amount of lived experience of socialism (none or little, several years or decades, most of the lifetime). The entire group of six women and three men comprised a student working part-time, a freelancer, three employees, and a pensioner.

Whenever any of the interviewees is cited in the subsequent sections, his or her stakeholder 'identity' and provenance are indicated accordingly.[4] Yet, it must be underlined that in many cases the stakeholder distinctions were not clear cut. For instance, two of the interviewed planners used to be engaged in urban activism before 'stepping to the other side'. Moreover, nearly all activists, planners, decision–makers, and entrepreneurs inhabit the cities under investigation, which qualifies them to speak from the perspective of citizens as well. In fact, the interviewees were fully aware of this double-agency and, at times, made clear which stance they were taking while discussing particular issues.

After the field research had been completed, the transcripts of digitally recorded interviews were coded and analysed with the use of MAXQDA software according to the procedures of grounded theory. The obtained results were compared and contrasted between the groups of respondents, and between the three cities. While an overview of the general findings is presented in the following section, the subsequent one comprises a more detailed discussion relating to the nine selected concrete narratives.

The post-socialist urban common good unpacked

The first part of the conducted interviews yielded comparable results among the identified categories of stakeholders in all the three cities. While the conversations usually began with a broad question about the interviewee's personal **definition of urban common good**, responses varied greatly across the groups. Unsurprisingly, these provided by decision-makers, planners and activists were the most detailed and compound, usually entailing urban space and additional ingredients: public (and to some extent private) property, common resources (such as greenery, unpolluted air, and water supplies), as well as spatial order, social norms and standards, interrelations, and a sense of community. In general, planners tended to be the most cautious to include elements of the private sphere, while activists eagerly equated the idea of urban common good with that of the city as a commons. The definitions offered by several respondents emphasised the duality of the term, distinguishing between its substantive and procedural interpretations. One of the planners (LDZ01) noted that the 'old-school' approach to common good in Polish spatial planning consisted in negotiating between public and private interests, while more recently there has been more attention paid to the urban common good as something which should not serve any particular interests but, as much as it is possible, everyone.

In contrast to the 'experts', citizens and entrepreneurs found this question the most challenging. The few interpretations presented by the former group were in majority intuitive, simplistic, and often restricted to public space:

> Some sort of shared spaces, like parks or playgrounds, I don't know what else…, some kind of centres where people can meet. Something like that, I think…
>
> (Young Citizen, GDK28)

Overall, the provided opinions demonstrate that the understanding of urban common good has still been in the making. Several respondents even reflected on its literal meaning, as if it were a totally new concept to them:

> I hadn't heard that phrase before, common good, but it is a cool phrase, because good already means something is good, and common also sounds cool because it is then for everyone.
>
> (Middle-aged Citizen, GDK39)

> The notion itself (...) is already a positive message, we hear that something is good, this way or the other, and also common. OK, so [the latter] may be associated a bit worse, but those two words put together sound cool.
>
> (Entrepreneur, KRK35)

Many interviewees across the interviewed groups highlighted the abstractness of the term. Some considered this to be a neutral feature, while the outlook of others was more critical. For instance, as pointed out in this rather cynical declaration of a decision-maker from Gdańsk (31), the unlimited capacity of the concept of 'urban common good' renders it meaningless:

> It is one of the phrases I really 'adore', because it means nothing. (...) It is a buzzword, a master key which fits different locks, and may be used in different ways.

The opinions on theoretical aspects of urban common good were next developed and complemented with the respondents' notions of **how the idea is put into practice**. This part of research revealed especially deep divisions between the groups and a considerable lack of mutual recognition and trust. For instance, almost all planners and decision-makers admitted that even if within their own milieu the common good was recognised as a leading professional principle, its implementation encountered numerous obstacles, such as lack of goodwill and/or silo mentality within urban institutions, financial constraints, or flawed legislation. However, quite many of them also blamed the 'individualistic Polish society' (Decision-maker, LDZ22)—especially citizens opting *for* rights and *out of* responsibilities—, as well activists with allegedly hidden political agendas. These tensions are well illustrated in the following tirade delivered by a Łodź-based decision-maker (32):

> And the trouble with urban movements is their mythical approach to governance, [which they think is dependent on] some 'them' who sit on huge chests of gold, who are bad dwarfs, not willing to share with anyone, and who spend this gold on silly things and monuments, when they could furbish the whole city instead. Of course, after several [activists] had found employment in [municipal institutions], all of a sudden, their viewpoint changed completely!

At the same time, planners and decision-makers alike perceived entrepreneurs as a group expected to choose pursuit of own interests over following the principles of common good. This stance met with disapproval from activists as much too forbearing and, in consequence, working against the common good. The entrepreneurs themselves displayed a notably pragmatic approach, taking advantage of the existing lax rules and regulations but also acknowledging that abiding by the principle of common good simply pays off:

> We don't ask ourselves what is common good, we have zoning plans [to comply with].
>
> (Entrepreneur, GDK42)

> I have noticed that [other entrepreneurs] begin to realize and understand that [respecting common good] works in favour of the corporate image and is also well received by local governments and public administration. And so, arranging some of the[ir privately owned] space in order to make it accessible [to everyone] works absolutely to their advantage.
>
> (Entrepreneur, GDK23)

If the citizens came out as a group the least interested in discussing the practicalities of urban common good, the activists tended to be the most disenchanted with the effectiveness of the term and did not consider it to be 'the subject of debates, nor [a slogan] mobilising people' (Activist, KRK21). Nonetheless, regardless of the divisions between the groups, two opinions were shared unanimously. Firstly, all the interviewees agreed that the reconciliation of the interests of various urban actors for the sake of common good was just as challenging as it was necessary. As one respondent put it, 'when we talk about common good, (…) we talk more of a certain ideal or utopia, an objective which we should be pursuing' (Entrepreneur, LDZ07). Secondly, despite the overall critical outlook on the idea's translation into practice, there was a consensus that relevance of urban common good within the local agenda had improved over the past decade.

In terms of **perceptions of commonness of urban space**, opinions did not vary substantially across the interviewed groups. The division line was usually drawn between 'the private' and 'the public'. Emblematic for the former were gated estates and enclosure, in general, was mentioned as the key distinguishing feature, more significant than property relations. Public space, on the other hand, was associated with openness and accessibility. Its other commonly specified attributes included inappropriate management, degradation, and the condition of 'belonging to everyone, that is, to no one really' (Decision-maker, GDK31). Likewise, the term 'common space' was used as a synonym for either public and 'no one's' space or un-classified, vague, residual space, unprotected and thus particularly susceptible to appropriation. Reference to common space understood as collective or shared could be traced mainly in the accounts of middle-aged or older respondents, who recalled that there had been much more of it in the socialist period, even if it had not been properly taken care of. The same group also agreed on the detrimental effects of the formerly imposed ideology of commonality. Namely, they claimed that it may have instigated Poles' aversion to urban common good, and induced glorification of the 'sacred right to property' after 1989. Yet, the transition of Polish cities from one extreme (forced commonality) to another (unrestricted individualism) was also identified as a potential stimulus for a social awakening, and eventual recognition of the right to the city:

> [Under communism] we had awfully missed private property, and unfortunately during the transformation period private property became a sacred thing. That has led to many absurdities, and now we are in this moment in time when we start to think: 'Hello! Where has this taken us?!'
> (Activist, KRK14)

> In the Polish People's Republic benches had been placed everywhere, [...] and grandmothers sat on them, and fed pigeons, keeping an eye on the grandchildren running around, but [after 1989] all at once there came this idea that the benches should be removed [...] because that was where the alcohol was being drunk, etc., [...] and then again [20 years later] these benches have started to reappear in the streets because it turned out that together with [the comeback of] the benches the elderly tend to return to public space [...]. I am under the impression that the last 20 years, i.e., the systemic change, have caused the society to forget what the city is all about.
> (Planner, LDZ04)

A few respondents questioned the validity of the term 'common space', either insisting on its obscurity or claiming that every space in the city must have a legal owner. The attitudes towards the private–public dichotomy revealed similar ideological differences across all the interviewed groups. For example, among the entrepreneurs, the opinions on the optimal type of management of urban common good varied from 'the belief that common space needs a manager is a harmful stereotype' (KRK11) to 'for the sake of the common space someone has to take care of it, and for someone to take care of it, financial resources are required' (GDK25). These differences were the most pronounced in the interviewees' outlooks on the ongoing **conflicts over the urban common good**. The citizens were the only group to find this question challenging. The other respondents indicated the issues listed below as the most frequent and/or noticeable in the three cities:

- urban transport and communication systems (domination of individual car transport vs. sustainable solutions),
- management of green infrastructure (urban development vs. protection of green areas),
- appropriation of public space through gating, advertising, etc. (economic interests vs. accessibility and spatial order),
- NIMBY attitudes towards urban space (individual vs. collective interests),
- touristification (tourists' needs vs. citizens' needs).

As noted by one of the planners (LDZ03), the most intense conflicts arise from the clash between two extreme visions of urban governance—'the free-market approach', represented by the private capital, and postulating minimum number of regulations, and the 'socially focused approach', exhibited by urban movements demanding maximum control and intervention. A corresponding dualism appears in a half-mocking comparison made by an activist from the same city:

> [another activist] aptly described this [situation, saying] that in Łódź there are two parties: the Quality of Life Party and the Golden Calf Party, and the front line is drawn between the two.
>
> (LDZ30)

Conversely, a majority of the interviewed citizens claimed to stay out of conflicts over urban space unless they themselves were involved. Only one representative of this group declared direct engagement in the fight for common good outside her own neighbourhood. Together with the activists, the citizens were, however, the most sensitive to various kinds of **exclusions from the urban common goods**, and were keen to quote their numerous workings:

commercialisation of public space (affecting the poor, incl. the elderly, and social tenants, but also limiting everyone's access to public space which becomes increasingly dominated by costly services),
barriers for pedestrians and prioritisation of private transport (disregarding the special needs of the disabled, the elderly and young parents),
selective deprivation of rights to public consultation and participation (experienced by students, immigrants, and other unregistered residents).

The marginalisation of 'uninfluential groups' whose needs tend to be ignored (e.g., pedestrians, who, unlike car-owners or cyclists, do not form lobbies or advocacy groups) was added to the list by several activists. Moreover, the activists turned out to be the most self-aware in terms of existing class differences and self-representation, depicting themselves as a 'middle-class companionship, birds of a feather flocking together' (LDZ41) but also pointing to the general middle-class privatist rather than communal orientation (Activist, LDZ42). The planners and decision-makers, in general, manifested less concern for exclusions taking place in urban space. Some of them would name a few discriminated groups but quite many either asserted that exclusions did not happen at all (Decision-maker, KRK43) or applied to those who were excluded 'of their own choosing', meaning the homeless or the 'unconcerned' (Planners LDZ04 and LDZ06, Decision-maker GDK31).

Finally, the interviewees were inquired about the **conditions which undermine urban common good** and, *vice versa*, these which favour it. As for the former, they included the popularity of gating, behaviour patterns inherited from the socialist period and the absence of culture of the common good, anti-communal bias in decision-making (not mentioned by any of the decision-makers but frequently brought up by the other actors), improper legislation (reported mainly by the planners and decision-makers), supremacy of private property rights, as well as attitudes of greed and self-interest. Furthermore, the activists indicated unequal power positions of certain social groups, the citizens focused on their limited involvement in decision-making, and the planners were the most concerned about potentially detrimental effects of citizen protests and NIMBY attitudes. The activists, citizens, and decision-makers alike agreed the lack of mutual trust and inability to cooperate among different urban stakeholders had negative impact. A consistent thread running through many of the accounts featured the Polish society as yet democratically immature but in the process of learning:

> I think this is just a certain [developmental] stage, I just don't know whether it's already teenage rebellion or still a three-year-old's rebellion, but we must experience all the childhood maladies to learn how to benefit from common good, and become responsible for it.
>
> (Activist, GDK19)

This reflection was supported by the ranking of **factors fostering urban common good**—a vast majority of the interviewees pointed to an increasing social awareness and sensitivity in that matter, along with a growing sense of agency, which translates into a sense of responsibility for, and identification with, common good. As pointed out by one of the planners (LDZ01), the previous overdose of privatism has caused the recent retreat into thinking of the city as a common good, even if it remains 'minority thinking' at the moment. Another planner (KRK13) spoke of 'an increasing trend' of sensitivity to common good, linking it to the society becoming progressively more affluent, which enables people to look, and act, beyond their basic needs. Other popular answers included social interactions and common initiatives, proactive local policies aimed at safeguarding common good in a long-term perspective, and improved legislation. In addition, the planners and activists emphasised the role of diffusion of best practices, and the entrepreneurs spoke highly of the positive impact of the private sector in terms of socially responsible place-making investment. The citizens were the most eager to praise the educational role of participatory budgeting.

Overall, regardless of the city, representatives of the five categories of respondents had generally similar reflections on urban common good, communicated similar problems, and employed similar narratives. The differences between the findings obtained in Gdańsk, Kraków, and Łódź applied mostly to the identification and prioritisation of the key threats to urban common good. In Gdańsk a too liberal approach to urban planning was pinpointed as the primary challenge, resulting not only in unsustainable development and spatial chaos but also leading to multilateral conflicts between the city makers (the local authorities, developers) and the city users (the residents, commuters). The interviewees in Kraków emphasised weak civic engagement and strong conservatism in decision-making coupled with the private property rule pushed to the extreme. Łódź, in turn, was perceived by the local respondents as a withered city struggling to reclaim its identity and citizen pride, which in their opinion explains a relatively low sense of commonality. For instance, as maintained by a representative of the planners (LDZ06), it is something which sets Łódź apart from Kraków, which did not experience comparable population losses after 1939. Another planner (LDZ04) attributes the same feature to the urban layout and the lack of a central public space.

The results of this part of the study thus confirm that urban common good in contemporary Poland has been, as of yet, either a notion in the (re)making, meaning various things to various users, or an empty signifier, meaning nothing and everything at the same time. Regardless of the socio-economic differences between the three case-study cities, the understanding and application of the term were similar among the interviewed groups of urban stakeholders. The decision-makers, the planners and the activists dealt with the term in a professional manner. The entrepreneurs had learnt to employ it to their own ends. By contrast, the citizens were rather unacquainted with the concept and tended to guess its meaning instead of providing their own definitions. Almost all the respondents highlighted the abstractness of 'urban common good' and 'common space', yet their reservations dissolved whenever the focus of conversation switched to such related issues as conflicts over urban space or exclusion from the right to the city. The hybridity of the term surfaced from most of the accounts, its meaning as a value however being more pronounced than the one of a resource.

With only some exceptions, the differences between the stakeholders' viewpoints seemed to be implacable—as if the cake of urban common good had been irrevocably sliced between separate interest groups. Therefore, at present the idea of urban common good seems to divide rather than unite, and does not entail community interaction. The interviews exhibited manifold variants of the 'us' versus 'them' dichotomy, together with a common tendency to shift the blame and responsibility on 'others' (e.g., the bad dwarfs). It was particularly noticeable in the opposing statements formulated by the planners and decision-makers against the activists, and *vice versa*. Between these groups, the concept of urban common good serves as a double-edged weapon, used both for defence and attack.

In light of these findings, the amalgam of the urban common good, or the contested gold from the bad dwarves metaphor, has been formed in the process of a continuous strife. While the respondents generally refrained from relating to the legacy of the socialist period, they readily emphasised the significance of the neoliberal agenda, and the supremacy of private property rights, recognising them as the driving forces behind the post-socialist urban change. Accordingly, what follows from their accounts is that the process of devaluation of urban common good initiated in socialism not only found continuation but was reinforced in the early decades of the transition. After the new system had settled in, the escalating clash of interests between urban stakeholders led to an exacerbation of conflicts and divisions. Only then, a discussion over urban common good and the process of its reinvention would begin, breeding further dissent but also providing space and potential for building the common ground across the antagonistic groups.

'Going back to the obvious'?: the forging of urban common good in concrete narratives

As it had been intended, an investigation of issues which combine into an inventory of concrete narratives allowed to expand the analysis and articulate more distinct threads within the post-socialist experience of urban commonality. I will first examine them separately for each case-study city, and then proceed to a synthetic discussion.

Gdańsk: lessons of urban commonality learnt the hard way

One of the most prominent themes transpiring from the Gdańsk interviews concerned the sense of 'lost' commonness. This aspect turned out to be especially complex in the instance of the post-shipyard grounds. On the one hand, almost all respondents agreed that following the process of privatisation, the area no longer holds the status of a collective urban resource, and remains common only in the symbolic dimension. On the other one—they quite unanimously perceive this very privatisation as a mistake which led to extensive demolition and commodification of a previously undervalued common good. According to a local activist, the site was 'appropriated, sold out and bereaved of the communal character' by means of a 'ruthless' market mechanism (GDK20). Correspondingly, one of the district councillors calls the effects of the ongoing transformations as 'wasted space' (GDK18). Another local decision-maker, identifying himself as a steadfast believer in sacred private property (GDK31), acknowledged that the former shipyard grounds should have been communalised first, comprehensively planned and only then made available for development to private parties.

Regardless of the fact that the municipality has never had much control over the area, representatives of several groups voice some sort of expectation that the local authorities assume liability for its current redevelopment, and thus make up for the botched zoning plan. One of the city planners refutes this claim on the grounds that the same zoning plan regulations may accommodate both 'good' and 'bad' investment (GDK29). From a yet another perspective, an independent planner (GDK44) identified multiple groups which over the years have come to claim their 'right to the shipyard'—in his opinion, apart from the legal owners guided by the idea of profit-making, they included artists, who were invited to domesticate the space and overstayed their welcome, the local activists pursuing an 'unknown agenda', and, finally, local entrepreneurs taking root through business establishments. Yet, according to the respondent, none of the said parties thought of the space in terms of common good, that is, outside of their own collective interests, trying to force their own narratives instead.

The only category of the interviewees not entirely recognising the site as a common good were the citizens. Among the few who did, there was a visible lack of sense of agency, manifest in their resignation from even the right to fantasise about the design of the Young Town. A retired employee, who admitted to being emotionally attached to the place, restrained herself from describing her own idea for the site's redevelopment, stating that she 'shouldn't be too much of a know-it-all' (Elderly Citizen, GDK27). Another citizen self-censored his vision of the place as too dreamy and unrealistic (Middle-aged Citizen, GDK39). One of the interviewed activists (GDK19) attributed the feeling of being left out to underwhelming citizen engagement in the fight for common good, dominated by the other groups, which she laments, considering citizens' bottom-up pressure as the one which could make all the difference.

This relative apathy of urban citizens is also observed in responsibility issues connected to the courtyards in the Main Town. In this case, all the groups empathise with the difficult position of the residents, acknowledging their preference for control over the space, even if its semi-public character renders the counterclaims of other potential users legitimate. However, what generates controversy, is the residents' wish for the yards to be enclosed and properly managed, while the wish is contradicted by their reluctance to buy or lease the space in question, which is the option advocated by the municipality. The figure of the manager is recalled in several accounts, accompanying the assumption that only ownership of space guarantees that it is properly taken care of, which evidently echoes the narrative of the sacredness of the right to property.[5]

Yet, the municipal officers' calls for restoration of the once-in-place distinction between private and public space (Planner, GDK29) are seen by some respondents as a necessary evil, which could only be an interim solution to the problem:

> Fencing these courtyards may be compared to an antibiotic treatment—if we are ill, it is obviously better to treat ourselves with natural remedies, like herbs and tea with lemon and honey, but if the fever is too high, we go to the doctor and take an antibiotic, sometimes by injection. And when we get better, we may take better care of ourselves by drinking herbs and tea, and eating well. The situation we have today is so critical that we can no longer rely on herbs and soft measures but need an injection. Fencing works like such an injection, clearly defining to whom the space belongs and who is responsible for it. After we do this, the residents will get the sense that they are in fact the managers of these spaces (…). When they learn to take care of them, and understand that the grass that grows there will have to be mowed, and when this knowledge is passed on to their children and so on, in 20–30 years these yards may open up again. Because there will be this awareness that these are semi-private spaces of the inhabitants of these tenement houses.
>
> (Activist, GDK19)

> Of course, I imagine such an ideal situation someday that this space becomes recognized as private and other people respect it, but for today I think that [fencing] is the only way to instil some order (…) so here I am siding with the inhabitants.
>
> (Activist, GDK24)

This process of learning, dubbed 'courtyard democracy' by the first of the quoted activists, is reported to have been extremely effective for the residents' empowerment. In the process, the neighbours not only had the occasion to express, confront and negotiate their ideas but also got to know each other better. Altogether, she thinks of these additional results of the programme as of the first step towards building common good at a microscale, which may later be upscaled to the local and supralocal levels.

Another take on the issue of residential gating as a way of managing common space is found in the respondents' explanations of the phenomenon. While the planners tend to justify it by the existing legal regulations, which allow enclosure of what is in fact private space, the activists express their disapproval of such practices, but at the same time show understanding for their complex motives. Even though Garret Hardin's approach seems to prevail over Elinor Ostrom's in all groups, there is decidedly less acceptance for gating in the inner city where its infringement upon commonality of space is greater extent than in the suburbs. The awareness of this problematic aspect of gating, based on the respondents' own experiences, including the usually less vocal citizens, seems to be on the rise. Discussing a potential ban on residential gating in cities, an activist (GDK19) related to the history of Gdańsk as a living urban organism taking nourishment from openness and commonality, which are the obvious truisms we should now be going back to. Two of the interviewed entrepreneurs demonstrated how the developers in Gdańsk already retreat from gating, having noticed that the tide is turning.

The overarching conclusion from the 'case studies within a case study' conducted in Gdańsk is that although the paradigm of the city as a commodity holds, the recognition of urban commonality has been progressing through a challenging, at times painful, process of learning from loss. The three examples show that regardless of the actual ownership status, and with considerable delay, the urban space has been increasingly perceived as common. As the right to the city, championed by the activists, continues to compete with the sacred right to property, backed by the decision-makers and planners, the citizens and entrepreneurs sit on the fence and jump off only occasionally. The former group however does not seem indifferent to the issues of urban commonality, and so the lessons from the model of 'courtyard democracy' give some clue of the ways in which the model of the city as a commons could be reinforced in the nearest future.

Kraków: selective citizen mobilisation for urban common good

The common pattern linking the Kraków cases in terms of their recognition as common goods consists in the success of grassroots initiatives finding traction with the wider public. Both the anti-smog and pro-Zakrzówek campaigns were conceived and set into motion by small protest groups, which eventually managed to effectively mobilise hundreds, if not thousands, of citizens. There is a broad consensus across the interviewed groups that while the problem of air quality has not been new in Krakow, it certainly gained notice in the last several years. The fact that the fight for unpolluted air was recognised as 'a common cause, which should be dealt with together, not individually' (Entrepreneur B, KRK11) not only helped to create a critical mass but even made the issue enter Cracovians' everyday conversations:

> Actually, this (…) is something that concerns and unites everyone. And it is about improving everyone's wellbeing. (…) This topic fades away in the summer, and always comes back like a boomerang in wintertime. (…). It's a kind of a taxi small-talk topic, as in 'So how is the smog today?'.
>
> (Young Citizen, KRK15)

According to the interviewees, another factor which largely contributed to the public support in the commoning of the quality of air is that collective action towards this goal was 'depoliticised', that is, devoid of associations with any political parties or agendas. An activist who had been personally engaged in the campaign particularly underlined this aspect:

> we used to repeat and still repeat very often, that we are not connected to any political party, that in this fight for clean air we do not pursue any individual goals, but that we do what we do for the common good (…) And I think that we have built this trust by showing that we were fighting for the common good (…) not to win political points, but because we want our children to breathe clean air, and the voivode to breathe clean air, and the chair of the city council or the marshal, and this is what I think was crucial in bringing our group together.
>
> (KRK26)

Conflicts over urban greenery in Kraków are reported to usually concern undeveloped areas performing the role of informal open green spaces. Whenever plans for their eventual development are made public, regardless of their lawfulness, they are perceived by citizens as appropriation of common urban space, which should be blocked. One of the decision-makers (KRK37) apprehends the citizens' attachment but finds their expectations unfounded in the light of the applicable law. A similar impasse is encountered in the case of numerous enclosed church gardens—despite their formal ownership, they are considered 'spaces lost for the community', access to which 'is only provided through Google Maps' (planner, KRK10).

Several side threads which came up in the interviews indicate the faultlines along which the opinions of different types of stakeholders divide. For instance, both in reference to air pollution and conflicts over urban green spaces, the activists devoted much time to raising the questions of social and environmental justice, whereas these angles were missing from the accounts of the decision-makers and planners. The citizens were less keen to identify with pro-environmental initiatives failing to comply with their own interests—one respondent complained that he 'felt stupid' and 'judged' when using his car but at the same time highly approved of protests over residential development on the fringe of a landscape park which he frequents for recreation (Middle-aged Citizen, KRK16). Another citizen admitted that she only got interested in collective action for urban greening after her children had been born and the use of open green spaces became a priority in her household. The saving of the Zakrzówek area, popular with regular residents, scuba divers, and rock climbers, is another case where 'it was relatively easy to ignite citizen involvement' (Planner, KRK13).

The question of direct involvement also shaped the respondents' insights on touristification. Some of the decision-makers and planners, as well as the citizens living in more peripheral neighbourhoods, did not see the inflow of tourists and the related effects on the quality of life as a serious problem, or anything more than a 'necessary evil' (Planner, KRK10), and pointed to the economic advantages which would presumably translate into general common good (Middle-aged Citizen, KRK36, Decision-maker, KRK 43). The opinions differed among the entrepreneurs—while the majority downplayed the issue, claiming that the situation used to be worse (Entrepreneur B, KRK11) and that tourism is only one of the elements of the local life (Entrepreneur, KRK12), one held a firm opinion that the residents were not only annoyed by the hordes of tourists but physically displaced from the inner city, parts of which 'are no longer habitable' (Entrepreneur, KRK35). A sense of symbolic displacement was recalled by one of the citizens to whom the formerly familiar Main Square turned into 'an alien space' (Yong Citizen, KRK15), while a district councillor (KRK17) quoted statistics of people who had moved out from the commercialised city centre over the preceding decade. Yet, the issue tends to have less galvanising potential than air pollution and urban green spaces enclosures:

> People do feel that something is not right, and that urban policy prioritizes the needs of tourists over the needs of residents, but this does not translate into any political mobilization.
>
> (Activist, KRK21)

Altogether, the examples from Kraków reveal the power of bottom-up pressure in reclaiming the right to the city, which, however, turns out to be effective only when several conditions are fulfilled. Firstly, the commonness of the urban feature must be recognised by the general public, and not just by activists or the most affected groups. Secondly, even though the city as a commons is a highly political concept, the urban stakeholders pushing for it refrain from any political connotations, in particular of partisan character. This is most probably connected to high levels of distrust in the so-called 'party politics', regarded as opaque and corrupt, but also to the societal divisions mentioned in Chapter 3. Taken together, these conditions, therefore, confirm the legitimacy of the concrete narrative approach in the Polish context.

Łódź: foot-in-the-door commoning of urban regeneration

In Łódź, the three narratives recount the changing approach to urban regeneration but also reveal how the stakeholders in the urban common good follow different rationales in attaining it. Starting from the City of Tenements programme, its role as 'the primary impulse for transformation of urban space' (Planner, LDZ01) is recognised by planners and activists alike. Both the groups emphasise that its objectives were mainly image-related but at the same time note how, owing to the aesthetic upgrade of the city centre, the citizens began to rediscover the city's heritage and build their local identity on its basis, and how it had mobilised private owners to renovate their properties, thus multiplying the effect of the municipal undertaking. However, its shortcomings—relating to displacement of tenants, a low technical quality of renovations, and generally revanchist inclinations—were ruthlessly exposed by the activists, some of the citizens, and the entrepreneurs.

The image which emerges from the assessment of the programme provided by these groups is that of an epitome of the city as a commodity framework—the accounts revolve around the issue of distributive spatial justice, and the 'whose right to the city' question is reflected on by a few respondents. Conscious of this critique, the city representatives carefully referred to it as 'a regeneration proving ground' (Planner, LDZ03), and underlined how their strategy had changed since, and focused on maintaining the social mix. One of the planners (LDZ40) repudiates the allegation of discrimination against municipal tenants who were denied return to the renovated dwellings by presenting their claimed vandalism as a lack of concern for the common good. However, neither the interviewed decision-makers nor the planners addressed the issue of the tenements as a common good in more detail.

The fact that the large-scale project of The New Centre of Łódź was still under construction at the time of the interviews, made any discussion on its commonality somewhat theoretical and abstract. However, the respondents' viewpoints on what was planned in the area are indicative of tensions between the neoliberal take on spatial planning and the pro-communal response. The former approach, as in the case of the tenements, boils down to thinking of the newly created space as a common good because of its boosting effect on the image of Łódź and its citizens' sense of pride (Entrepreneur LDZ07). Yet, the protracted process of planning and implementation, whereby the vision had changed multiple times, rather exposed a lack of an overall conception and a disconnection from the actual needs and preferences of citizens. In addition, as specified by one of the planners (LDZ01), for quite a long time the consecutive redevelopment plans treated this area as if it was an island, isolated from the rest of the city—the idea for its integration with the adjacent neighbourhoods, and with the high street of Piotrkowska, occurred relatively recently.

As suggested by several interviewees, the explanation of this inconsistency lies in the project's reliance on the—eventually unsuccessful—bid for EXPO 2022, which they denote as only one out of a few examples proving how the spatial planning in the first decades of the 21st century hinged on the neoliberal agenda:

> Such a paradigm of neoliberal thinking was dominant—it was actually intended as a district of office buildings and flats, for, I don't know, people from Warsaw and commuting to work back there, or for companies that would move from Warsaw to rent offices here. (…). And the effect would be that on the one hand we would have Tuwima Street with dilapidated tenement houses, and on the other, a completely new multimillion investment simply draining resources and being this brave new world.
>
> (Planner, LDZ01)

> I went to a couple of meetings about the New Centre of Łódź. It is clearly a top-down idea of the local authorities, captivated by a vision of a super-elite ghetto in the city centre.
>
> (Activist, LDZ02)

A related charge, put forward mostly by the activists, concerns insufficient public participation in the planning process. For instance, a citizen who happened to take part in expert consultations, claims that the elaborated recommendations were not taken into consideration, which made her lose affiliation to this space (Older Citizen, LDZ08). Correspondingly, one of the activists refused to comment on the project at all, on the grounds that it was 'detached from the city and the public good' (Activist, LDZ30). This opinion is somewhat confirmed by the city planner's concession that the New Centre of Łódź programme had not been broadly consulted with the citizens, nor based on their expectations of the area's development, but it was 'more the issue of urban planners and architects' and their vision as the actual experts (LDZ03).

Citizen participation and agency finally transpired in the interviews when the conversations switched to the issue of woonerfs. Here, nearly all the respondents concurred that the innovation had become a symbol of a novel communal way of thinking of the city. Different groups, however, indicated different elements as key to the experiment's success. The activists particularly emphasised the participatory dimension and the fact that the project had been discussed with the inhabitants of the neighbourhood prior to being put together as a proposal but also drew attention to the integrative effect of the process on the local community. The planners and decision-makers talked mainly of civic education and taking over responsibility by citizens, at the same time admitting to being surprised how the potentially controversial project, in terms of its limitation of traffic in the city centre, caught on. The other outcomes reported by the respondents as having a pro-communal impact, involved 'communication egalitarianism', that is, equal treatment of pedestrians, cyclists, and drivers (Entrepreneur, LDZ07), the development of a sense of belonging and domestication of public space (Planner, LDZ06, Activist, LDZ02), and the 'pretty balcony effect', which the inhabitants of the other neighbourhoods were quick to copy (Planner, LDZ04). Only among the citizens, the opinions ranged from apprehending the woonerf initiative as 'the essence of urban common good' (Young Citizen LDZ05) to hardly seeing it in such context (Middle-aged Citizen, LDZ08).

The revolutionary change of approach to urban regeneration in Łódź, instilled through the woonerf initiative, clearly breaks away from the paradigm of the city as a commodity. Interestingly, despite the local authorities' long attachment to the latter, the new model seems to have been acclaimed by the decision-makers and planners alike. As summed up by one of the city councillors (LDZ33), 'the woonerfs were forced by the citizens—and good for them!'. In a more illustrative manner, the resulting embrace of participatory standards by the city officials was also compared to 'getting one's foot in the door' (City planner, LDZ04). While it should not be assumed that the new paradigm entirely replaced the dominating neoliberal mode of city-making, the empowerment of the citizens and a change in the narrative come up in almost every account.

Between commodification and commoning of the city: findings from Gdansk, Kraków, and Łódź

The study of concrete narratives in the three case-study cities reveals a plethora of interesting threads which could be followed in further research. What I would like to point to is the general dethronement, or demonopolisation, of the city as a commodity paradigm. The ways in which urban stakeholders conceptualise and practice urban commonality may vary substantially but, regardless of these differences, it is clear that the emerging alternative of the city as a commons has been acknowledged, or at least recognised, across all the interviewed groups. The commoning of the city is still at a relatively burgeoning stage but while there is no guarantee that it will flourish, it has already made a difference in Gdańsk, Kraków, and Łódź.

While the general pattern of the city as a commons paradigm being forced bottom-up against the top-down city as a commodity agenda applies to all the three cities under investigation, some specifics may be observed in each of them. In Gdańsk, the revaluation of urban commonality arrived too late for some of the valuable commons to be saved from irreversible commodification, but the ongoing lessons in urban citizenship hold potential for the future. The spectacular achievements of collective action for the urban common good in Kraków relied on wide citizen support, which in turn was highly dependent on the alliance of individual and collective interests.

Quite contrastingly, in Łódź the embracement of the city as a commons paradigm happened via an inconspicuous initiative whose unforeseeable success turned the tables by modifying the approach to urban regeneration.

Another finding following from the analysis concerns the differing comprehension of urban commonality between the groups of urban stakeholders. It should be recalled here that personal biographies of quite many respondents involved two or even more stakeholder identities, which sometimes blurred their affiliations with a touch of ambiguity. However, they would usually be very specific about the perspective from which they were talking. The quest for the urban common good turned out to be fundamental for the activists, who tend to build their identity on the fight for the right to the city. They are also fully aware of their game-changing impact. The planners and decision-makers seem to be currently learning to accommodate the notion of urban common good, which to some extent results from the pressure instilled by urban movements but also from general global trends. Opinions within the latter group often varied depending on whether I talked to an incumbent deputy mayor or an opposition councillor. The attitude toward urban common good was the most heterogeneous among citizens—some of them took it in intuitively, some others did not identify with it at all or were only beginning to be aware of it. The entrepreneurs fell somewhere between the four other groups and were the most inclined to refer to their personal experience as the citizens, signalling it accordingly.

In general, this part of the research brought more specific outcomes than those obtained in response to the more abstract questions on the definition of urban common good or perception of commonality of urban space. Even the citizens who had offered modest answers during the introductory part of the interviews would turn more loquacious when asked directly about specific local issues. Therefore, the adoption of the concept of a concrete narrative for research purposes yielded productive results. Yet, as much as it was possible to hold interesting conversations with all of my interviewees, some of them would point out that it was the first time they were actually reflecting on common good in the urban context. In the words of one of them, the interview made him realise that this topic was in fact foreign to him, and changed his previously more negative outlook on urban commonality in Kraków (Entrepreneur, KRK12).

Insufficient citizen engagement and civic education, the dictate of the private property and the neoliberal programme, the domination of the middle-class agenda, the absence of culture of common good, and meagre social solidarity were the recurring issues in the respondents' accounts. Yet, what especially drew my attention was the much-exposed thread concerning the question of political neutrality. Again, the concept of concrete narrative is helpful here as almost all interviewees, with very few exceptions, evaded the use of the terms 'politics' or 'political'. As mentioned earlier, this probably has to do with the level of trust and the deep socio-political divisions in the Polish society. This however bears specific consequences, as, for instance, the activists who decide to run in local elections, often risk losing credibility, and are generally met with scepticism. The stance on political neutrality was explicitly criticised by only one respondent. In his opinion, 'it is the most foolish thing in Poland that we had assumed that politicians are kind of aliens, because whoever does not participate in political life is simply powerless' (Decision-maker, LDZ32).

A few issues touched upon by the respondents leave opportunity for a more in-depth research. One such matter relates to the question of scale, often exceeding the city boundaries (cf. Kip 2015). For instance, when discussing the issue of smog in Kraków, a few respondents noted that the scope of the problem is supralocal, and its systemic solution necessitates measures to be taken at the supralocal scale—regional, in terms of capping pollution in the neighbouring gminas, but also national, in terms of adequate legislation. Another aspect concerns the interplay of individual and collective interests.

While I found little direct evidence for positive impact of the socialist experience on the current rediscovery of urban common good, it was looming in the background of numerous accounts. As I did not inquire about it directly, it may have remained overlooked or unrecognised by some respondents. Yet, the question posed in the title of this chapter, using a quote from one of the interviewees, remains only partly unanswered—overall, the respondents' accounts suggest that Polish cities' return to the previously lost 'obvious' value of commonality hinges on an informed liaison between the more distant and more recent past and present.

Notes

1 The number of inhabitants in Gdańsk is relatively lower but comparable to those in Kraków or Łódź provided the inclusion of the adjacent cities of Gdynia and Sopot—with Gdańsk they form the so-called Tricity. Treating them together is justified by the spatial continuity of the three urban centres and strong functional links between them.

2 Although this statement is generally true, it is important to acknowledge several aftershocks following from the political turmoil in Europe, especially at the time of Napoleonic Wars. **Gdańsk**, incorporated into the Kingdom of Prussia following the II Partition of Poland in 1793, was dependent on the Napoleonic France between 1807 and 1814 as the Free City of Danzig. After it had returned to Prussia (the latter succeeded by the German Empire in 1871 and the Weimar Republic in 1918), it became a semi-autonomous Free City again in 1920, maintaining this status until WWII. The annexation of **Kraków** by Austria took place in 1795 (III Partition), yet in 1809, the city briefly joined the ephemeral Duchy of Warsaw created by Napoleon two years before. The Congress of Vienna reversed this decision in 1815, at the same time, establishing the Free City of Kraków. The failure of the Kraków Uprising in 1846 put an end to the city's partial autonomy and fully restored the Austrian administration until 1918. **Łódź** first fell under the Prussian rule in 1793 (II Partition) but only 13 years later, in 1807, was incorporated into the Duchy of Warsaw, and in 1815, formed part of Congress Poland, subjected to the Russian Empire. It then remained under Russian control until the beginning of WWI, when it was taken over by the German Empire. As of 1918, only Kraków and Łódź were included in the re-established Republic of Poland (the Second Polish Republic), Gdańsk joined them no sooner than in 1945.

3 One of the interviews featured two respondents (entrepreneurs) at the same time, at their own preferences, explained by the wish for equal gender representation. The 28:17 ratio of male to female respondents in the sample reflects gender inequality in access to institutions of money and power in Poland—females were more numerous than males only in the activist and citizen groups.
4 The numbers (01–44) accompanying the three-letter acronyms of the case-study cities represent the order in which the interviews were conducted. For instance, 'Middle-aged Citizen, GDK39' stands for a representative of inhabitants of Gdańsk, aged between 40 and 59, with whom I held the 39th interview out of the total 44.
5 The most radical exemplification of this neoliberal line of thought is found in a decision-maker's reference to the US legislation, which allows the use of arms as a means of protecting private property from trespassers.

References

Brzeziński K (2011) Nowe centrum jako remedium na problemy Łodzi. *Przegląd Socjologiczny*, 60(2–3), 393–422.

Dankowska M (2016) Śródmiejski krajobraz kulturowy w dokumentach planistycznych, na przykładzie strefy wielkomiejskiej w Łodzi. *Studia KPZK*, 168, 51–76.

Flyvbjerg B (2006) Five misunderstandings about case-study research. *Qualitative Inquiry*, 12(2), 219–245.

Frankowski J (2020) Attention: smog alert! Citizen engagement for clean air and its consequences for fuel poverty in Poland. *Energy and Buildings*, 207, 109525. https://doi.org/10.1016/j.enbuild.2019.109525

Gądecki J (2012) *I ♥ NH: Gentryfikacja starej części Nowej Huty?* Warszawa: Wydawnictwo IFiS PAN.

Gdańsk 2030 Plus Development Strategy (2014) Gdańsk: Urząd Miejski w Gdańsku. Retrieved from: https://www.gdansk.pl/strategia-rozwoju-miasta-gdansk-2030/gdansk-2030-plus-development-strategy-download,a,38041

Gdańskie Nieruchomości (2022). Retrieved from https://www.nieruchomoscigda.pl/podworka/projekty/ZPGM (last access: 11 February 2022).

Glejt-Uziębło P, Uziębło P (2018) *Partycypacja w Trójmieście: O prawnej regulacji mechanizmów demokracji semibezpośredniej w Gdańsku, Gdyni i Sopocie*. Gdańsk: Wydawnictwo Uniwersytetu Gdańskiego.

Gorzelak G (2006) Poland's regional policy and disparities in the Polish space. *Regional and Local Studies*, Special Issue: 39–74.

Grabiński T (2015) Rozmiary ruchu turystycznego w Krakowie w latach 2010–2015. In: Borkowski K (ed.), *Ruch turystyczny w Krakowie w 2015 roku*. Kraków: Małopolska Organizacja Turystyczna, 49–53.

Grabkowska M, Makowska L (2011) Mała solidarność mieszkańców: nowy fenomen Gdańska? In: Michałowski L (ed), *Gdański fenomen: próba naukowej interpretacji*. Warszawa: Wydawnictwo Naukowe Scholar, 192–203.

Grabkowska M, Szmytkowska M (2021) Gating as Exclusionary Commoning in a Post-socialist City. *Region*, 8(1), 15–32.

Häikiö L (2007) Expertise, Representation and the Common Good: Grounds for Legitimacy in the Urban Governance Network. *Urban Studies*, 44(11), 2147–2162.

Hrehorowicz-Gaber H (2013) Effects of transformations in the urban structure on the quality of life of city residents in the context of recreation. *Bulletin of Geography. Socio-economic Series*, 21, 61–68. https://doi.org/10.2478/bog-2013-0021

Immigrant Integration Model (2016) Gdańsk: Gdańsk City Hall. Retrieved from: https://download.cloudgdansk.pl/gdansk-pl/d/20170691579/immigrant-integration-model.pdf

Jakóbczyk-Gryszkiewicz J (ed.) (2015) *Procesy gentryfikacji w obszarach śródmiejskich wielkich miast na przykładzie Warszawy, Łodzi i Gdańska. Studia KPZK, 165*, Warszawa: KPZK PAN.

Janik L (2020) Rola Zakrzówka jako elementu sieci zieleni miejskiej Krakowa. *Problemy Rozwoju Miast*, 68, 91–99.

Jarzębiński M (2014) Inwestycje infrastrukturalne w Krakowie w kontekście przygotowań do organizacji EURO 2012. *Problemy Rozwoju Miast*, 1, 61–77.

Jażdżewska I (ed) (2010) *Duże i średnie miasta polskie w okresie transformacji. XXII Konwersatorium Wiedzy o Mieście*. Łódź: Wydawnictwo Uniwersytetu Łódzkiego.

Kajdanek K (2017) Narracje i sfery dyskusji o mieście. In: Bierwiaczonek K, Dymnicka M, Kajdanek K, Nawrocki T (eds) *Miasto, przestrzeń, tożsamość - studium trzech miast: Gdańsk, Gliwice, Wrocław*. Warszawa: Wydawnictwo Naukowe Scholar, 155–181.

Kamrowska-Załuska D, Kostrzewska M (2014) Wielkie wydarzenia jako katalizator procesów rewitalizacji. *Problemy Rozwoju Miast*, 2, 57–70.

Karsten L (2009) From a top-down to a bottom-up urban discourse: (re)constructing the city in a family-inclusive way. *Journal of Housing and the Built Environment*, 24(3), 317–329.

Karwacka G (2010) The influence of international sport events on development of infrastructure: Euro 2012 in Poland – case study. *Romanian Review on Political Geography*, 12(2), 375–385.

Khan S, Van Wynsberghe W (2008) Cultivating the under-mined: Cross-case analysis as knowledge mobilization. *Forum Qualitative Sozialforschung*, 9(1): 34. Retrieved from http://nbn-resolving.de/urn:nbn:de:0114-fqs0801348 (last accessed: 30 May 2017).

Kip M (2015) In: Dellenbaugh M, Kip M, Bieniok M, Müller AK, Schwegmann M (eds) *Urban Commons: Moving Beyond State and Market*. Moving beyond the city: Conceptualizing urban commons from a critical urban studies perspective, Basel: Birkhäuser, 42–59.

Klich-Kluczewska B (2005) *Przez dziurkę od klucza. Życie prywatne w Krakowie (1945–1989)*. Warszawa: Wydawnictwo TRIO.

Kostrzewska M, Rembeza M (2012). European Football Championship-Euro 2012 in Poland: The impact on the host-cities' development. *Journal of US-China Public Administration*, 9(10), 1215–22.

Kraków Development Strategy: This is where I want to live. Kraków 2030 (2018) Kraków: Municipality of Kraków. Retrieved from: https://www.bip.krakow.pl/zalaczniki/dokumenty/n/225388/karta

Kruczek Z (2018) Turyści vs. mieszkańcy. Wpływ nadmiernej frekwencji turystów na proces gentryfikacji miast historycznych na przykładzie Krakowa. *Turystyka Kulturowa*, 3, 29–41.

Krzymiński D (2015) Znikająca stocznia i jej obrońcy: wędrówka subiektywna trasą miejsc (nie)istniejących. In: Mendel M (ed.) *Miasto jak wspólny pokój: Gdańskie modi co-vivendi*. Gdańsk: Gdańskie Towarzystwo Naukowe, 71–98.

Kuchcik M, Milewski P (2018) Zanieczyszczenie powietrza w Polsce – stan, przyczyny i skutki. *Studia KPZK*, 182(2), 341–364.

Kubicki P (2011) Nowi mieszczanie – nowi aktorzy na miejskiej scenie. *Przegląd Socjologiczny*, 60(2–3), 203–227.

Lamprecht M (2016) Ewolucja kwartałów śródmiejskich Łodzi w kontekście kurczenia się miasta. Współczesne wyzwania. *Studia Miejskie*, 23, 99–115.
Liszewski S, Young C (eds) (1997) *A comparative study of Łódź and Manchester: geographies of European cities in transition. Łódź*: University of Łódź.
Lorens P, Bugalski Ł (2021) Reshaping the Gdańsk Shipyard—The Birthplace of the Solidarity Movement. The complexity of adaptive reuse in the heritage context. *Sustainability*, 13, 7183. https://doi.org/10.3390/su13137183
Madejska M (2014) Łódź – studium przypadku. In: Bukowiecki Ł, Obarska M, Stańczyk X (eds), *Miasto na żądanie. Aktywizm, polityki miejskie, doświadczenia.* Warszawa: Wydawnictwa Uniwersytetu Warszawskiego, 227–237.
Marcińczak S, Musterd S, Stępniak M (2012) Where the Grass is Greener: Social Segregation in Three Major Polish Cities at the Beginning of the 21st Century. *European Urban and Regional Studies*, 19(4), 383–403.
Mazurkiewicz M (2021) The games that never happened: Social reception and press coverage of the Kraków Bid for the 2022 Winter Olympics (2012–2014). *The International Journal of the History of Sport*, 38(13–14), 1350–1368. https://doi.org/10.1080/09523367.2021.1997998
Mergler L, Pobłocki K, Wudarski M (2013) *Anty-Bezradnik przestrzenny: prawo do miasta w działaniu*. Warszawa: Fundacja Res Publica im. H. Krzeczkowskiego.
Model for Equal Treatment (2018) Gdańsk: Gdańsk City Hall. Retrieved from: https://download.cloudgdansk.pl/gdansk-pl/d/202009155785/broszura_model_na_rzecz_rownego_traktowania_2020.pdf
Murphy E, Fox-Rogers L (2015). Perceptions of the Common Good in Planning. *Cities*, 42, 231–241.
Murzyn MA (2006) *Kazimierz. The Central European experience of urban regeneration*. Kraków: Międzynarodowe Centrum Kultury.
Murzyn-Kupisz M (2012) Przemiany historycznych dzielnic śródmiejskich w dobie neoliberalnego urbanizmu. Przykład Starego Podgórza w Krakowie. In: Szmytkowska M, Sagan I (eds), *Miasto w dobie neoliberalnego urbanizmu*. Gdańsk: Wydawnictwo Uniwersytetu Gdańskiego, 83–105.
Peisert A (2013) Partycypacja obywatelska jako przejaw odpowiedzialności za dobro wspólne: zróżnicowanie strukturalne i regionalne. In: Geisler R (ed.), *Odpowiedzialność – przestrzeń lokalnego społeczeństwa obywatelskiego, biznesu i polityki*. Opole: Instytut Socjologii Uniwersytet Opolski, 62–75.
Peräkylä A (2009) Analiza rozmów i tekstów. In: Denzin NK, Lincoln YS (eds.), *Metody badań jakościowych*, vol. 2, Warszawa: Wydawnictwo Naukowe PWN, 325–343.
Perkowski P (2013) *Gdańsk – miasto od nowa: kształtowanie społeczeństwa i warunki bytowe w latach 1945-1970*. Gdańsk: Słowo/Obraz Terytoria.
Polanska DV (2010) The emergence of gated communities in post-communist Poland. *Journal of Housing and the Built Environment*, 25(3), 295–312.
Prawelska-Skrzypek G, Morgan A (2020) The return to Europe or the return to solidarity? GdańskShipyard—case study in organizational culture. *Sustainability* 12, 7032. https://doi.org/10.3390/su12177032
Program Rewitalizacji Łodzi 2026+. Gminny Program Rewitalizacji (2016) Uchwała Nr XXXV/916/16 Rady Miejskiej w Łodzi z dnia 28 września 2016 r. w sprawie przyjęcia Gminnego Programu Rewitalizacji miasta Łodzi. Łódź: Urząd Miasta Łodzi.
Program Rewitalizacji Łodzi 2026+. Gminny Program Rewitalizacji (2018) Uchwała Nr LXXIII/1980/18 Rady Miejskiej w Łodzi z dnia 5 lipca 2018 r. zmieniająca uchwałę w sprawie przyjęcia Gminnego Programu Rewitalizacji miasta Łodzi. Łódź: Urząd Miasta Łodzi.

Purchla J (2018) *Miasto i polityka: przypadki Krakowa*. Kraków: Universitas.
Rzeńca A, Sobola A (2018) Budżet obywatelski jako instrument kształtowania przestrzeni Miasta–przykład Łodzi i Katowic. *Biuletyn KPZK*, 272, 205–215.
Śmiechowski K, Burski J (2018) Transition: The post-industrial orphan in neoliberal Poland 1989-1994. In: *From Cotton and Smoke: Łódź – Industrial City and Discourses of Asynchronous Modernity 1897–1994*. Łódź-Kraków: Wydawnictwo Uniwersytetu Łódzkiego.
Strategia Rozwoju Gdańska do roku 2015 (2004) Uchwała Rady Miasta Gdańska Nr XXXIII/1011/04 z dnia 22 grudnia 2004 r. Gdańsk: City Council of Gdańsk.
Strategy of Integrated Development of Łódź 2020+ (2012) Łódź: Urząd Miasta Łodzi.
Szczepańska A, Zagroba M, Pietrzyk K (2021) Participatory budgeting as a method for improving public spaces in major polish cities. *Social Indicators Research*. https://doi.org/10.1007/s11205-021-02831-3
Szechlicka J, Kamrowska-Załuska D, Mrozek P, Szustakiewicz J (2016) Non-places in the centre of the historic Main Town in Gdansk? – Design Thinking as a method of solving problems in cities. *Biuletyn Komitetu Przestrzennego Zagospodarowania Kraju PAN*, 264, 306–324.
Szmytkowska M, Nowicka K (2015) Neo-liberal reality in post-industrial waterfronts of the post-socialist cities: The polish tri-city case. *Economic and Business Review*, 17(2). https://doi.org/10.15458/85451.10
Szpakowska-Loranc E, Matusik A (2020) Łódź – Towards a resilient city. *Cities*, 107, 102936. https://doi.org/10.1016/j.cities.2020.102936
Tobiasz-Lis P (2016) Łódź "od-nowa" – o odnowie miasta z perspektywy codziennego doświadczania jego przestrzeni. *Studia Ekonomiczne Regionu Łódzkiego*, 21, 77–86.
Tracz M, Semczuk M (2018) Wpływ turystyki na zmianę funkcji przestrzeni miejskiej na przykładzie Krakowa. *Biuletyn KPZK*, 272, 272–284.
Uchwała Krajobrazowa Gdańska (2018) Uchwała Nr XLVIII/1465/18 Rady Miasta Gdańska z dnia 22 lutego 2018 r. w sprawie ustalenia zasad i warunków sytuowania obiektów małej architektury, tablic reklamowych i urządzeń reklamowych oraz ogrodzeń, ich gabarytów, standardów jakościowych oraz rodzajów materiałów budowlanych, z jakich mogą być wykonane, na terenie Miasta Gdańska. Gdańsk: Urząd Miejski w Gdańsku.
Uchwała w sprawie utworzenia parku kulturowego pod nazwą Park Kulturowy Stare Miasto (2010) Uchwała nr CXV/1547/10 Rady Miasta Krakowa z dnia 3 listopada 2010 r. Kraków: Urząd Miasta Krakowa.
Zdyb M (2017) Proces rewitalizacji a jakość życia mieszkańców – projekty Zielone Polesie i woonerfy w Łodzi. *Space-Society-Economy*, 21, 73–97. http://dx.doi.org/10.18778/1733-3180.21.04

5 (Re)making the urban common good in post-socialist cities

Several years ago, in an interview about the upcoming Polish-edition of a well-known book on urbanity, *The Seduction of Place*, its author—architectural historian Joseph Rykwert—explained that what he meant by likening the workings of cities to that of a dream was the dialectics of the conscious and the subconscious (Żakowska 2013). He next went on to say that cities are the result of the sum of the choices constantly made by different urban stakeholders. The latter include citizens, whose even smallest interventions—such as the planting of a tree—add to the overall effect. This way, Rykwert unknowingly related to the bed-time story which I reference in the introduction to this book. Correspondingly, the multi-threaded narration developed through the four following chapters inevitably brings us to a moment in which final conclusions regarding the relation between urban common good and the post-socialist city must be drawn.

Ever since the ancient Greek concept of the polis, the notion that the city is not neutral but politically charged and socially construed has pervaded Western European thought. The works of Henri Lefebvre and numerous other scholars build on it to revolve around urban commonality and the related notion of urban common good. The latter can be conceptualised in many ways. One of the general distinctions is between the procedural and substantive meanings—of a value or principle and a shared resource, respectively. The model based on the definition of public good by Robert Asen (2017), which I introduced in Chapter 1, reconciles both strands. One of its components, the paradigm of the city dominant in a given public sphere, is relationally dependent on collective urban actors and the dominant rationales they pursue.

Throughout time, paradigm shifts occur, marking critical transitions in the socio-political making of urban commonality. In Western Europe, such a threshold was most recently reached when the neoliberal city as a commodity formula, dominant since the 1980s, was questioned and undermined by urban movements' demands for the right to the city. A few decades after 1989, a similar response in the form of the paradigmatic city as a commons emerged in countries east of the former Iron Curtain. However, as outlined in Chapter 2, the legacy of state socialism and intensity of transition to market capitalism, paired with additional conditions reaching pre-socialist times, rendered this process quite distinct in CEE.

DOI: 10.4324/9781003089766-8

The transformation of approaches to the production of urban common good—formerly at the centre of the state's ideology, but unsupported by socio-cultural standards and consequently devalued, entirely rejected after 1989 due to promotion of individualism and privatism, and finally rediscovered in the 2010s—offers a fascinating subject for research. Regardless of existing specifics among the states in the region, CEE cities after socialism followed a similar sequence of paradigm change. It first involved departure from the idea of the city as municipal infrastructure ('inherited' from socialism) and embracement of that of the city as a commodity (dominating during the transition to capitalism). Afterwards, in response to the latter, a competing concept of the city as a common good emerged and has been developing since. An overview of specific examples of urban change documents how this transformation impacted the critical spheres of urban commonality—housing, public transport, green infrastructure, public spaces, urban regeneration, and spatial justice.

The overall conclusion following from Part I of the book is that the failed socialist experiment of reducing the city to a communal infrastructure, devoid of the element of agency by reason of authoritarian control, heavily affected the formation of urban citizenship in the second half of the 20th and the first two decades of the 21st century. Moreover, the ideology-driven repudiation of private property rights not only did not translate into primacy of collective over individual interests but quite the reverse, equating anything 'common' with 'no-one's'. It also contributed to all the more uncritical adoption of the neoliberal agenda following the systemic change and the resultant, extensive commodification of the urban and the city. Disempowered by the dictate of the state and disillusioned with its performance during the socialist era, urban citizens across CEE entrusted city development to market forces, inadvertently moving from one extreme to another. Next, as the model of the city as a commodity struggled to live up to their expectations, urban movements took the initiative in lobbying for the right to the city and promoting an alternative vision of the city as a commons. This vision in fact relies on the return to the idea of the city as a communal infrastructure enhanced by the missing element of citizen participation in decision-making. As much as this configuration should not be perceived as a utopian ideal nor a panacea for all contemporary urban challenges, out of the three propositions, it is certainly the most inclusive and empowering with respect to urban citizens.

Part II of the book offers a more detailed insight into the current career of the urban common good in Poland. The evident recent expansion of the concept is found to partly resemble parallel Western European patterns and partly derive from the conditions underpinning the combined socialist and post-socialist experiences of urbanity. This general finding is based on the outcomes of two empirical studies—examining the changing approach to the urban common good in the legal, media and academic discourses after 1989, and exploring apprehension of the same notion by various types of urban stakeholders interviewed in Gdańsk, Kraków, and Łódź. Even though cities from Western Europe were often recalled as benchmark models in both of them, the dark and the light sides of the more and less distant past—of socialism and transition to capitalism, respectively—turned out to be equally influential.

Outcomes of the discourse analysis undertaken in Chapter 3 demonstrate that the gradual revaluation of the formerly compromised urban common good was inspired bottom-up and incidental rather than effecting from deliberate policies or strategies. The appraisal of urban commonality—fostered by urban movements and making its way into the mainstream since around 2011—was first communicated through print media. The evolution of topics relating the concept of common good to various urban issues reveals how they gathered gravitas over time. Initially, the analysed press articles mostly channelled debates on the tragic condition of neglected and/or commodified public spaces, only a few years later switching to reporting innovative practices of urban commoning—including experiments in participatory governance. While academic discourse has recently caught up with ongoing social change, legislative framework tended to lag behind—30 years on from the passing of the Act on Local Self-Government, urban citizens' standing within Sherry Arnstein's (1969) ladder of participation in decision-making still does not reach beyond the middle rungs of consultation and placation. Consequently, regardless of the narrative of the right to the city being increasingly broadcast in the public sphere, citizens' voices have remained relatively marginalised and unheard.

The investigation of accounts given by citizens, activists, decision-makers, planners, and local entrepreneurs featured in Chapter 4 allows us to draw three basic conclusions. Firstly, the interviewees' attitudes towards the urban common good seem to be likewise influenced by the knowledge of good and bad practices from the socialist era, and the experience-based awareness of neoliberal processes negatively affecting the urban common good. Secondly, although conflicts of interests between stakeholders in the urban arenas are inevitable, understandings of the urban common good across the five groups of representatives appear to be quite consistent, and their objectives and motivations are less irreconcilable than is usually assumed. Thirdly, along with the maturation of the paradigm of the city as a commons, the interests of various groups have not as much aligned as the differences between them have become acknowledged, leaving room for the potential future departure from antagonistic in the direction of agonistic relationships.

Regardless of varying local contexts, mechanisms of the ongoing (re)construction of urban common good in the three case-study cities appear to be very much alike. In Gdańsk, the absence of such an agenda at the beginning of the 2000s significantly reduced the potential for redevelopment of the former shipyard area as an inclusive and accessible urban space. Yet, the belated protest reactions of local residents and activists sparked enough citizen concern for the urban common good for the decision-making processes to become slightly reorientated towards wider citizen participation. Attempts at reconciliation of different stakeholder positions with priority given to the real needs of citizens are visible in the handling of the issue of management of courtyards. Correspondingly, bottom-up pressures in the other two cities—initiatives against air pollution and the fragmentation of open green spaces in Kraków, as well as participatory undertakings in Łódź—instilled gradual changes in top-down policies, but also publicised and popularised the narrative of the right to the city. This does not mean that the idea of commonality has become fully embraced by all urban stakeholders, nor that it has escaped being exploited to serve rather particular interests. Nevertheless, the research findings confirm that even though its standing is still fragile, the urban common good has become more than a buzzword in all three case-study cities.

On the other hand, what also follows from the results of the empirical research is that 30 years after the fall of the previous regime the quality of governance at the local level in Poland is highly unsatisfactory. Although decision-making processes are much more citizen-oriented than in the previous system, they are still far cry from inclusionary and integrative. Different groups of urban actors compete on unequal terms, dependent on their power position, failing to communicate with each other. For instance, citizens' interests are oftentimes better served by activists than by their legitimate representatives, that is, democratically elected decision-makers, but activists' ideas of the actual citizens' needs may also be detached from reality. Moreover, middle-class interests appear to be relatively over-represented. Therefore, even if the increased civic engagement, bottom-up pressure on local agendas and diffusion of participatory initiatives altogether combine to create a crawling development of the culture of common good, access to control over the city as a commons remains uneven. This condition renders the levelling of the field for less powerful urban stakeholders and more effective involvement of the so far marginalised groups in local decision-making processes a matter of key importance.

From this point of view, more mediation and constructive debates, along with the provision of equal opportunities on the extremely discriminative 'power ladder', should encourage reconciliation of seemingly disparate interests of various urban stakeholders or at least enable the identification of possible middle grounds. Yet, such a programme may be not easy to implement since—as evidenced both across the analysed discourses and in the interviewees' accounts—urban conflicts tend to be perceived as undesired, even pathological, rather than holding the potential for pluralist negotiation of clashing interests and perspectives and thus fuelling reinvention of the urban common good. To some extent, this follows from the inability of distinguishing the common good from political interests, and a broader problem of distrust towards 'the political' in general, resulting in a preference for depoliticisation. Inherited from the previous era, the latter factor impedes the recasting of the post-socialist local community as *com-munis*, with its focus switched from forced consensus-building to responsibility-taking (Amin and Howell 2016: 11).

As demonstrated in Chapters 3 and 4, one effective way of overcoming the detrimental effects of political and world-view divisions at the local level is collective action for the common good compliant with the assumptions of a 'concrete narrative' (Mergler et al. 2013). This idea has been recently picked up and incorporated into the concept of 'effort democracy' put forward by Jakub Wygnański (2021). The point of departure of his proposition consists in the thesis that currently, the politically disengaged Polish society tends to be more often bound by negative energy of disapproval and protest rather than any constructive agenda for what they would like to achieve together. He, therefore, considers it imperative to 'find a method, space and energy to formulate common demands and dreams; a method of broad, common reflection on what is common' (18). To achieve that he insists on a cyclical process of deliberative democracy, which would consist of five steps: choosing a topic for debate—starting from the most across-the-board and least controversial, followed by diagnosing and mapping of the problem, looking for solutions, giving the floor to citizen deliberation, and formulating and implementation of the elaborated proposal.

This is just one of many possible measures which could be applied for the emerging paradigm of the city as a commons to break away from Hardin's fallacy of overexploitation and doom and approach Ostrom's well-governed ideal instead. However, there are also multiple challenges ahead of the urban common good other than insufficient citizen empowerment and engagement. I will point to three of them which I recognise as the most imminent in Poland, but also elsewhere in the region. The first one concerns the possibly divisive and anti-communal aftereffects of the COVID-19 pandemic. The second relates to the burdens taken on by local governments which following the Russian invasion of Ukraine had volunteered to become refuge cities and remained largely unassisted by the central administration. This pressure is further enhanced by the post-pandemic and post-invasion soaring inflation and rising costs of energy. The third challenge is connected to the looming threats of democratic backsliding and populism (Havlík 2019, Pixová 2020, Bernhard 2021). While the examination of each one of these threads is deserving of a separate book, I will leave them unattended and only hint at seeking possible guidance in Ash Amin' (2006) four Rs of urban solidarity—the principles of repair, relatedness, rights, and reenchantment—introduced in Chapter 1.

To end this book on a more optimistic note, I would like to assert that I strongly believe in the force of urban commonality. Even though challenges arise, cities persist. They withstand wars, floods, financial crises, and pandemics. The reason for it largely lies in the common efforts of those who inhabit them, and in their common dreaming.

References

Amin A (2006) The Good City. *Urban Studies*, 43(5/6), 1009–1023.

Amin A, Howell P (2016) Thinking the commons. In: Amin A, Howell P (eds), *Releasing the Commons: Rethinking the Futures of the Commons*. London and New York: Routledge, 1–17.

Asen R (2017) Neoliberalism, the Public Sphere, and a Public Good. *Quarterly Journal of Speech*, 103(4), 329–349. https://doi.org/10.1080/00335630.2017.1360507

Arnstein SR (1969) A Ladder of Citizen Participation. *Journal of the American Institute of Planners*, 35(4), 216–224. https://doi.org/10.1080/01944366908977225

Bernhard M (2021) Democratic Backsliding in Poland and Hungary. *Slavic Review*, 80(3), 585–607.

Havlík V (2019) Technocratic Populism and Political Illiberalism in Central Europe, *Problems of Post-Communism*, 66(6), 369–384. https://doi.org/10.1080/10758216.2019.1580590

Mergler L, Pobłocki K, Wudarski M (2013) *Anty-Bezradnik przestrzenny: prawo do miasta w działaniu*. Warszawa: Fundacja Res Publica im. H. Krzeczkowskiego.

Pixová M (2020) Contested Czech Cities: From Urban Grassroots to Pro-democratic Populism. *Geografie*, 118(3), 221–242.

Wygnański J (2021) *Inny pomysł na demokrację*. Warszawa: Laboratorium Więzi.

Żakowska M (2013) Pokusa miejsca: Interview with Joseph Rykwert. *Dwutygodnik.com*, 6(110). Retrieved from https://www.dwutygodnik.com/artykul/4611-pokusa-miejsca.html (last accessed: 2 May 2021).

Index

Note: Page numbers in *italics* refer to figures and **bold** refer to tables.

For Product Safety Concerns and Information please contact our EU representative GPSR@taylorandfrancis.com
Taylor & Francis Verlag GmbH, Kaufingerstraße 24, 80331 München, Germany

www.ingramcontent.com/pod-product-compliance
Ingram Content Group UK Ltd.
Pitfield, Milton Keynes, MK11 3LW, UK
UKHW020140200926
475702UK00005B/832
* 9 7 8 0 3 6 7 5 4 5 7 3 4 *